Conversations About Anthropology & Sociology

Conversations About

ANTHROPOLOGY
&
SOCIOLOGY

Edited by Howard Burton

R **Ideas Roadshow**
INTELLIGENT. INQUISITIVE. INTERNATIONAL.

Ideas Roadshow conversations present a wealth of candid insights from some of the world's leading experts, generated through a focused yet informal setting. They are explicitly designed to give non-specialists a uniquely accessible window into frontline research and scholarship that wouldn't otherwise be encountered through standard lectures and textbooks.

Over 100 Ideas Roadshow conversations have been held since our debut in 2012, covering a wide array of topics across the arts and sciences.

See www.ideasroadshow.com for a full listing of all titles.

ISBN: 978-1-77170-306-2 (hardcover)
ISBN: 978-1-77170-177-8 (paperback)
ISBN: 978-1-77170-176-1 (eBook)

Edited, with preface and all introductions written by Howard Burton.

All *Ideas Roadshow Conversations* use Canadian spelling.

Contents

ON ATHEISTS AND BONOBOS
A CONVERSATION WITH FRANS DE WAAL

EMBRACING THE ANTHROPOCENE
MANAGING HUMAN IMPACT
A CONVERSATION WITH MARK MASLIN

Textual Note

The contents of this book are based upon separate filmed conversations with Howard Burton and each of the five featured experts.

Fred Gitelman is a world-champion bridge player and co-founder of Bridge Base Online. This conversation occurred on April 5, 2014.

Ian Stewart is Emeritus Professor of Mathematics at the University of Warwick and a bestselling science and science fiction writer. This conversation occurred on March 8, 2013.

Joseph Curtin is an award-winning violinmaker, acoustician and MacArthur Fellowship winner. This conversation occurred on May 2, 2013.

Frans de Waal is the Charles Howard Candler Professor of Psychology at Emory University and the Director of the Living Links Center at Yerkes National Primate Research Center. This conversation occurred on April 23, 2013.

Mark Maslin is Professor of Geography at University College London. This conversation occurred on October 12, 2016.

Howard Burton is the creator and host of Ideas Roadshow and was Founding Executive Director of Perimeter Institute for Theoretical Physics.

Preface

I fully expect that many readers will be quite taken aback when they look at the list of people involved in Ideas Roadshow's Anthropology & Sociology collection.

"A bridge player? Mathematician? Violinmaker? These are, quite clearly, neither anthropologists nor sociologists! How is this possible?"

Frans de Waal and **Mark Maslin**, being respectively a primatologist and geographer might well be able to be classed as "anthropologists", but how can these other people be justified? Indeed, such confusion might be rendered even more acute when one considers that two Ideas Roadshow experts that unequivocally *can* be considered sociologists—Nick Couldry and Denis McQuail—are placed in the Language & Culture collection. What, on earth, is going on?

Well, in some ways what is going on is a deliberate attempt to force you to transcend the often somewhat arbitrary disciplinary boundaries and encourage you to confront, from first principles, what we actually *mean* by anthropology and sociology to begin with.

And once you do that, I'm convinced that you'll start to see the obvious relevance of choosing someone who represents mankind's longtime fascination—even, you might say, addiction—to games, not only from his extensive personal experience in high-level competition, but also as the founder of the world's most popular online bridge sites. Of course, if need be we could formally justify such an excursion into this aspect of "cultural anthropology" by throwing around references to Geertz's Balinese cockfights or Huizinga's *Homo Ludens*. But I prefer the more direct approach: pointing out that the intrinsic attraction for recreational activities is so strong that, even without consciously planning on doing so, its pursuit will inevitably be wedded with other conspicuously endorsed activities, such as societal acclamation and commercial success. As world-champion bridge player **Fred Gitelman** describes it:

> *"The idea that Bridge Base Online has become a very successful business is interesting, because it was never intended at all to be a business. Basically, one day Sheri and I decided that there should be a free, high-quality online bridge site, just because that would be good for the game. We had no idea at all that it would eventually become a successful business. But it grew and grew, and we got to the point that every day hundreds of thousands of people were logging in to our site. I think it's that case that if you have enough people involved in something, you can eventually find a way to turn it into something that makes money."*

And then there's mathematics. Surely, you might think, if there's one discipline that seems irrevocably distinct from anthropology and sociology it would be the austere realm of professional mathematics? Well, there might well be an argument there had our conversation with mathematician and bestselling author **Ian Stewart** been narrowly focused on technical aspects of his mathematical insights. But it is not. Instead, virtually all topics under discussion are oriented towards the extremely human activity of *doing* mathematical research, attempting to straighten out deep-seated societal misconceptions about what mathematics actually is, commenting on how so much of contemporary society deeply depends on mathematical understanding and the steady accumulation of mathematical knowledge and reflecting upon the many sociological barriers—such as gender discrimination—that we regularly put up to inhibit people from participating in the mathematical enterprise:

> *"It was very striking to me that there was clearly a very long historical tradition of women in science in Portugal. What is it? How did this happen? There is a Women in Mathematics organization and I think they understand this. But one of the problems for women in mathematics is that there are still not as many as there should be. I*

think it is improving, there are figures that show it is improving at certain levels.

"But if you are a very good woman mathematician, the demands on your time are absolutely enormous. For a start, you have to get involved in the Women in Mathematics organization! Whereas men don't have that problem.

"It's fairly clear that this is an interesting problem that needs to be investigated, this is something that is not as well known as it should be, it also goes counter to some people's prejudices, so they tend to dismiss it. But it is absolutely clear. I don't know why. But it's a fact."

Our last apparently "out man out" is **Joseph Curtin**, a renowned violinmaker and acoustician who has done pioneering work on probing our societal need to perpetuate the longstanding myth that the Stradivarius violin represented the acme of human accomplishment that could, somehow, never be duplicated.

"Unfortunately, I think there's been a general anti-science feeling in the violin-making world and even in the music world. I can't tell you how often I used to hear that there's a kind of pride in the mystery of the violin. Some people think, Why would you want to lose that mystery? Of course, that's not how it works: the more you understand, the more interesting something is.

"How many times do you read media headlines that say something like, For centuries, scientists have struggled in vain to understand the secrets of Stradivari... Really? In vain? Do you have any idea how many interesting things have been learned?"

Meanwhile, Emory University primatologist and bestselling author **Frans de Waal**, moves us in yet another direction, offering up valuable insight on what it is to be human by explicitly comparing us to our non-human fellow travellers, en route consistently exploding our unjustified assumptions of uniqueness:

"I used to think that monkeys would care about getting less than somebody else but they wouldn't care about getting more. But now we

find that the chimpanzees do care about getting more: sometimes they will refuse getting good food if the other one doesn't also get good food. So they also play the ultimatum game the way humans play it.

"So now if you ask me what the difference is in the sense of fairness between chimpanzees and humans, I don't know what the difference is. It's not clear to me anymore what the difference is."

But while Frans highlights our inherent connectedness to other members of the animal kingdom, it cannot be denied that our power to change our habitat not only goes vastly beyond other creatures, but also—even more significantly—anything that we previously possessed in the past. Which brings us directly to UCL geographer and climate scientist **Mark Maslin**, who focuses his attention on the very large-scale issue of the so-called Anthropocene: the notion that the human imprint on the globe has become so large that it is now necessary to completely overhaul our framework of how we look at both the natural sciences and the social sciences. Today we urgently need to integrate them so that we and all the other life forms on the planet are able to flourish into the distant future:

"For me, then, the big thing is that in the early 21st century we've now hit that choice of which direction to go. The big problem is: How do you organize seven and a half billion people who are used to working in small groups, because of our brain, of about 150? How do you make that step up from a small tribal creature, which we are, to a mega-creature that covers the whole planet? I think that's the challenge that we're going to have this century."

For all sorts of reasons, then, it's important to start thinking differently about things.

In the Cards

A conversation with Fred Gitelman

Introduction
The Art of Competition

Why do we play sports and games? Is it for love of the game or the thrill of competition?

Sports journalists are all too prone to dip into hoary metaphors of artistic expression to describe a well-struck ball or a graceful save, but what's behind the thousands of sweaty hours of muscle memory that enabled such acts of on-field accomplishment to begin with?

Are most professional competitors simply in love with the idea of competition—desperate to capitalize on whatever opportunities might be available to push themselves to the highest international level? Or does it somehow depend on the activity in question? Would Roger Federer have been a great soccer player if raised in Brazil? Would Michael Jordan?

And what about so-called mental athletes—those who opt to submit themselves to similar levels of training, sacrifice and psychological pressure in an attempt to attain international glory in a decidedly less physical, but no less stressful, arena?

Fred Gitelman serves as an interesting case in point. A professional bridge player who has won numerous high-level competitions, including a World Bridge Championship in 2010, he is also the creator of pioneering bridge educational software, the founder of the world's largest internet bridge-playing site, Bridge Base Online (BBO), and has been a personal bridge coach to the likes of Warren Buffett and Bill Gates.

Impressive stuff. But bridge? Isn't that just a game for old fogeys?

Well, no. According to Fred, the world's best bridge players tend to peak at age 40, when their combined emotional maturity and neural processing power are most closely aligned. The reason, he believes, that most people today think of bridge as an "old person's game" is a legacy of its precipitous decline since its heyday midway through the last century.

> *"Back in the 1950s, things were very different. If you didn't play bridge, you knew people who played. If you walk up to an average person today and ask, 'Do you know someone who plays bridge?' they might reply, 'Well, my grandmother plays bridge.' But of course, their grandmother was a young person in the 1950s when everybody played bridge."*

Nevertheless, despite its undeserved branding nightmare as a game only for retirees, bridge is still very much alive and kicking. It's played all over the world and increasingly online, in no small way thanks to Fred and his BBO colleagues. And even though their site is free to all, the company can still make a considerable profit by simply charging one dollar for its associated tournaments.

> *"Only about 10% of the people who log in to our site ever play these tournaments and give us a dollar. But there are so many people playing that 10% of them is quite a lot. It's very strange making money one dollar at a time, but if you have enough dollars it works."*

Interesting enough. But Fred's story is particularly intriguing because it gives us a glimpse into this fiercely competitive mental arena. What actually happens, I wondered, when the best bridge players in the world sit down to compete? What toll does that sort of competition take, mentally and physically?

> *"It certainly takes a lot out of you. The tournament I just played in Dallas was 10 days long. You play about eight hours a day in two sessions. And when you're playing, you're just sitting there thinking and thinking, intensely concentrating. At the end of the day, it doesn't matter what kind of shape you're in, you're exhausted. By the end of the week, I've usually lost about 10 pounds."*

Not much different than a professional athlete at all, then. And a rather far cry from the average grandmother.

But let's return to our question of motivation. What's behind it all? An insatiable lust for competition in any form? A testosterone-driven determination to humiliate one's opponent?

For Fred, at least, nothing could be further from the truth. As a teenager, he simply fell in love with the game and couldn't stop himself from devoting every available moment to improving his skills. It had nothing to do with competition per se, he maintains—indeed, his is of an inherently non-competitive nature. But then, at a certain point, he felt that he had no choice but to change his fundamental psychological outlook if he was going to continue improving.

> *"After I had been playing bridge for about 10 years, I began to get noticed by the better players in Canada, the elite players who represented Canada internationally, and they started to invite me to play bridge on their teams. At the time it was a massive honour: these people were my ultimate heroes. I wanted to be just like them and I was thrilled to have a chance to play with them. They were highly competitive people who were trying to be the best in the world, and if I wanted to stick around them I had to force myself to become much more competitive than I ordinarily would have been, because they were expecting that.*

> *"I guess, consciously or not, I underwent some kind of personality transition and became a competitive person. At that time I thought my goal was to become the best, and the best is to win the world championships. That was around 1990, so it took me 20 years to do it. But 20 years later, in 2010, I won a world championship."*

Successfully undergoing this sort of "competitiveness transplant" in order to become a world champion can't be too common. But what happens then, I wondered? Does the change become permanent? Or do the competitive juices fade away after having finally scaled the summit?

For Fred, it definitely looks like the second option has the upper hand for the moment.

> *"Well, the final chapter has not been written yet, but I think that's the way it's likely to go. It's kind of strange, really. I still love the game very much, but I'm just not that interested in playing competitively right now. I can't say for sure that that's how I'm going to feel in a year or two, but for now I feel like I've done what I set out to do. I'll stay very interested in bridge, of course, but I'm not necessarily trying to be the best."*

Score one for the love of the game, then. After having become world champion, that is.

The Conversation

I. Bridge Background

What it's all about

HB: I'd like to explore your personal story: how you got into bridge and what the life of a professional bridge player is like. But first, we should probably talk a bit about the game itself to provide some contextual background. I think it's best to assume that there's a fair number of people who don't know anything about it whatsoever. So I'm going to turn it over to you to give a top-level description of the game.

FG: Sure. Well, bridge is a card game. To play, you need four people and a deck of cards. It's considered to be a challenging game, in that one could spend a lifetime playing bridge, studying the game and competing, and one would never really master it. That's one of the great things about bridge: you can always get better at it. I've been completely serious about it since I was a teenager; and more than 30 years later, I still feel like I learn something every time I play. What that's really all about is that the number of ways that you can shuffle a deck of cards is comparable to the number of atoms in the universe, or something like that: it's a massive, massive number. And every way the cards are shuffled gives rise to a different bridge deal that corresponds to a different problem that can be solved—you're never going to see everything. You'll sometimes see variations on a theme. Some of the ways you'll shuffle the cards will result in a problem that is almost trivial to solve. Some will result in a problem that is almost impossible to solve. But most of the time, it's somewhere in between; and if you apply yourself and do everything right, you can figure it out.

I think that's probably what drew me to the game in the first place. It was the sense that this was a problem that I could figure out in the same way that people find crossword puzzles appealing, or sudokus, or Rubik's cubes. You get a certain satisfaction out of figuring out the answer and knowing that you've got it right. That's why bridge first attracted me as something I wanted to be good at.

HB: I'm going to get to that shortly, but let's first review the basics. So far all I know is that you need four people and a deck of cards.

FG: Well, it helps to have a table too. The four people sit around a table. It's a partnership game, with each pair sitting opposite each other as partners. You use the whole deck each time, all 52 cards, so everybody gets dealt 13 cards.

Now there are two phases to every bridge deal. The first thing is called "the bidding".

Bidding is done verbally. Everybody looks at his or her hand and describes the hand using a specific language to tell the partner information about his hand, such as, *"This is the number of hearts in my hand", "Here is the number of aces in my hand"*, and so forth. You're having a conversation with your partner, telling each other what you have; and eventually the idea is that you know enough about your partner's hand that you can make a statement about the combined strength of the two hands and what you would like to try to achieve in the second phase of the bridge deal.

This second phase is "the play of the cards". In the play of the cards, the cards are played out one at a time until all 52 cards have been played. Meanwhile, it's the bidding that determines the specific goal of the play of the cards, the contract that one team is trying to achieve during the play of the cards, while the other team is trying to stop them from achieving it.

HB: This bidding process is a public conversation, right?

FG: Correct.

HB: So it's not just that your partner hears what you're saying, but your opponents do as well.

FG: That's right. And one of the interesting things about bridge is that there are no secret codes. If you're serious about that game, you and your partner will spend a great deal of time discussing various bidding sequences and what the bidding sequence has to say about the nature of your hand for any particular bid.

But according to the rules, your opponents are always entitled to ask you, *"What did he mean when he bid that?"* and you have to tell them. Everything is very above-board in the bidding: whatever information you share with your partner, you also have to share with your opponents.

HB: So if I'm reading this and know nothing about the game, I might ask, *"Well, what's the point of that? If this is all public, why don't you just tell someone exactly what's in your hand? Why even go through this whole ritual with a separate bidding language?"*

FG: That's a good question. The nature of the language that's used for bidding is such that there's only a certain number of words in the language, and there are rules that determine which words you can use at which time. It happens to be the case that there aren't enough words to describe every possible bridge hand that you might have—not even close.

One of the things that's challenging about the game is that you only have some information about what your partner's hand looks like. You don't know exactly what his 13 cards are, and you have to estimate, based on the rough picture that you have, what you think the appropriate team goal should be. It would be a lot easier if your partner could just show you his hand and you would know exactly what you should be trying to do, but it doesn't work that way. You only get a rough idea.

HB: Right. Let's move on a bit now to the public image of the game. My sense is that most people now associate bridge with older people:

a recreational social game that's on par with canasta or something like that. That's the knee-jerk sense, I think, that people who don't play the game necessarily have.

That certainly wasn't the case for you when you got interested. And my sense is that there's nothing necessarily age-dependent about the game at all. Of course, there are many older people who play, but bridge is by no means a game just for older people.

FG: No, that's certainly true. When bridge was at its peak of popularity, which probably happened in the 1950s, it was a very different situation. If you didn't play bridge, you knew people who played. Now, millions of people who played in that generation are elderly, if they're still alive.

So if you walk up to an average person today and ask, *"Do you know someone who plays bridge?"* They might reply, *"Well, my grandmother plays bridge."* Of course, your grandmother was a young person in the 1950s when everybody played bridge. Nowadays, unfortunately, I'm still considered a relatively young bridge player, even though I'm almost 50 years old. That's really a function of the fact that the game hasn't been marketed particularly well to people of my generation or younger generations. While there are still some bridge players who are my age or younger, and some of them play extremely well, there aren't nearly as many as there are elderly people. That's something that I'd very much like to see change, but I have to admit I'm not particularly hopeful it's going to happen.

HB: I'd like to explore your pessimism in detail a little later on, but for now let's talk about the evolution of the game before getting to your personal story. I don't pretend to know very much about this, but my sense is that it developed from the card game whist. Is that right?

FG: Yes.

HB: And when did that start? Were people playing whist in medieval times? Do you have some sense of this?

FG: I have some sense of it, but it's not something I know a great deal about. As far as I can tell, the history before the early 20th century is not very well known. But there are various card games that go back to ancient times, and some of the predecessors of bridge, including whist, were played in Europe in the Middle Ages.

HB: Really?

FG: Yes. But bridge as it's played now is actually called "contract bridge". The rules of contract bridge were formalized in the mid 1920s. Before that there was another popular variation called "auction bridge", and that was a descendant of whist, which, I believe, goes back another couple of hundred years.

HB: What was the difference between contract bridge and auction bridge?

FG: Largely the bidding. The bidding in contract bridge was more sophisticated and I believe the scoring was different. Auction bridge basically doesn't exist anymore, but whist is still played in some circles; but while it used to be an extremely popular card game, it's now a niche activity.

HB: So another thought people may have is, *Since this is a card game, it necessarily depends on the draw of the cards; it depends on luck.* But there are ways of actually filtering the luck factor out, right?

FG: Well, let me start by saying that in its simplest form, when you start playing bridge socially with four people, there is certainly an element of luck that will determine who will win. I personally think that's one of the strengths of the game: there's just enough luck that if you're a competent player and things go your way, you have a chance to beat someone better than you. You don't have a particularly good chance, but you have a good enough chance that you'll keep coming back, because you'll know that you're not going to lose every single time. I think that's appealing.

When it's played competitively, when there are trophies or prize money on the line, you don't want the luck factor to be so important. So what's done in those circumstances is that, instead of just shuffling the cards and dealing and shuffling and continuing like that, once the cards have been shuffled they are preserved in something called a duplicate board.

This is called "duplicate bridge"; and the whole idea is that all 13 cards for each hand get put back into these slots, so that the entire hand gets preserved and passed along to another table to play after it's been played somewhere else.

This way we can calculate the score by how well you do compared to other people who have the same cards as you have. So it doesn't matter so much if you are dealt good cards or bad cards; what really matters is what you do with the cards compared to other people in the same circumstances.

Questions for Discussion:

1. Have you ever played bridge? Do you agree that it is generally regarded as "a game for old people"? Why do you think that is?

2. To what extent do you agree that the widespread popularity of bridge in the mid-20th century was a reflection of cultural values? If so, what values do you think those were, exactly?

II. A Personal Perspective

Beginnings, partnerships, tournaments and memory

HB: And now, finally, we get to your story. You became captivated by bridge as a young man, right? How did that happen for you, and did your bridge interest make you unusual?

FG: Well, I was a pretty unusual kid to begin with, I must admit. I grew up in the 70s and 80s and was the first kid on my block to own a personal computer. So I was fairly well-known amongst my school-mates as a weird, geeky kid who played with computers.

I happened to have three friends who had been taught by their parents to play, and they needed to find a fourth person to fill out their bridge game. They thought, *Fred's this kid who likes playing with computers, maybe he'll like playing bridge.* So they approached me.

You might say that I'm a bit of an obsessive-compulsive type, so as soon as I discovered the game, I went to the library to take out every book I could on bridge. I would frequent the bridge clubs in Toronto where I grew up, sitting for hours watching the best players play and asking them questions about why they did what they did. It didn't do much for my standard education, I must admit: I was supposed to be going to high school at the time.

HB: How old were you then?

FG: 16 or 17.

HB: What were the next steps? You got hooked by the game after being exposed to it by your friends, and then you threw yourself into it whole hog. Where did you go from there? I presume you got better and better and better.

FG: Right. It's like any other activity that you can play competitively: there are various levels that you can play at. The first level of competitive bridge is the local club, where every afternoon and every evening they'll have 15 or 20 tables of people all playing bridge and there's nothing really at stake. It's just about trying to beat the other people in the club.

That's a step up from playing socially, but it's not super-serious. I started by doing that. After that there are local tournaments. Maybe once a month there's a tournament in Toronto, or 20 minutes outside of Toronto, where you can win a little trophy. Then there are regional tournaments where a couple times a year I would drive to Montreal, Detroit or Buffalo; and if you won you would get your name in a bridge magazine. Beyond that, there are national championships; and beyond that, if you're successful, you can represent your country in the world championships.

You might think that that would've been my goal: to just keep winning bigger and bigger tournaments. But I didn't particularly care for that so much. I liked to go to these tournaments because I liked to play against better players.

HB: And challenge yourself.

FG: Exactly. And meet the people I had read about who were my heroes and learn from them. But winning the tournament in and of itself was not an important goal for me. I was not particularly naturally competitive. I just thought it was a beautiful game and I wanted to learn everything I could about it.

HB: And did you find, as you developed, that you had natural skills in some areas as opposed to others? You spoke just now about the difference in the game between the bidding and the playing. Did you have a sense that your real forte or natural ability was in a certain area, or that you had to work harder on this or that?

FG: I was naturally drawn to the play of the cards. It may be because that's something that's almost purely technical, that you could become good at if you studied it independently. It's something that you can do on your own if you read enough books and put in enough time, whereas the bidding is something that you really have to master with your partner. It isn't that I was completely anti-social or didn't want to work with partners, but rather, because I was so obsessed with bridge and wanted to work on it 24 hours a day, it was hard to find someone like me. So a lot of the work I did was individual by nature, and that lent itself to becoming proficient at the play of the cards.

HB: Because you could control your development there a bit easier.

FG: That's right.

HB: I'd like to talk a bit more about partnerships. We've mentioned that bridge is a partnership game, but when you just talked about progressing through different levels: playing at your club, playing in local tournaments, regional tournaments, national tournaments—well, that is the type of thing that you'd imagine a tennis player or a golf player would do. Yet those are solo activities, whereas in bridge you're always playing with a partner. Do you have a difficult time finding people? Do you choose different people all the time? How does that work?

FG: If you're really serious, you'll have basically one regular partner. It's almost like a marriage: you and that person are a team, a bridge team, and you'll spend a great deal of time together. After you play you'll discuss all the deals that came up and say things to each other like, *"You bid this on this deal; what were you trying to do? I didn't understand."* You begin to learn your partner's habits, just like a marriage. You begin to appreciate the emotional side of things too—you learn to recognize when your partner's having a bad day and learn ways to make him feel better, ways to compensate for his not being at his best.

HB: Right. But when you're starting off, you don't have that luxury.

FG: No. When I was starting off I would just go to the bridge club and hope that there would be someone there who would be my partner. I got a little bit lucky in that regard because it was unusual for someone so young to go to the bridge club, and the regulars there were concerned about me because they knew I should have been in school. But they were very welcoming: they liked it to see a young person playing bridge. And I was so interested in learning and was well-mannered and respectful. I really wanted to know what other people thought and I asked their opinions and listened to them. Of course, people naturally like it when you do that, because it gives them credit for knowing something and makes them feel good.

HB: And shows that you are willing to learn from them.

FG: Yes. Now, if you were to just go to play in a bridge club without a partner, there would be a decent chance that the manager would be able to find someone else for you to play with who didn't have a partner either. But it has to be said that sometimes these people don't have a partner for a reason: because they're not very polite or they don't play very well. But generally, if you're someone who behaves himself, or has reasonable bridge skills, or ideally both, you don't have that much trouble finding people to be your partner.

HB: I'm guessing you've seen a fair number of couples playing bridge to the detriment of their relationship.

FG: Yes.

HB: You hear these horror stories of personal relationships going wrong because of playing bridge, and either they'll have to stop playing bridge or stop seeing each other. Or perhaps both.

FG: You're absolutely right.

HB: Are there combined marriage/bridge therapy sessions on offer?

FG: I would probably have to say that most married couples would be better off not playing bridge together. The reason is that it becomes an ego battle for a lot of couples. The truth is, any bridge partner has to manage the conflicting egos. The nature of the game is such that sometimes you're going to get bad results no matter how well you play and it can be frustrating. Sometimes you just get unlucky and your opponents play very well and you'll do poorly because your opponents play well. It can be frustrating and upsetting, and it's natural for many people in these circumstances to just blame someone other than themselves. Your partner is often a built-in person to blame. So it's often the case that, if the couple doesn't have a completely ideal relationship, they'll look for ways to blame each other and take out their frustrations in real-life on their partner at the bridge table: they'll lash out. You often see that with married couples.

HB: You sometimes play with your wife, though.

FG: I do. And perhaps this sounds terrible, but my wife Sheri would be the first to tell you that one of the ways we make it work is because the ego battles don't come into play. She's a very fine player, but she knows that I am a much better player, a world-class player. So if we have a bridge disagreement—

HB: She'll defer to you.

FG: That's right: she'll defer to me. Of course, Sheri deserves a lot of credit here: she's a good student who's able to keep her ego in check and doesn't always have to prove that she's right. I probably deserve some credit as well, because I know all too well that nobody's perfect at this game; it's too difficult. So if I make a mistake, I'm not above saying, *"Sorry, Sheri, I screwed that up."* We're able to manage that situation pretty well.

HB: But when you play at the highest international levels, you have, as you were talking about before, a set partner. I guess it would be somewhat analogous to a tennis player playing professional doubles, where you play for different periods of time with different partners. Do you think about prospective partners all the time? As you travel from tournament to tournament, do you say to yourself, *"Gosh, that guy would actually fit with me, if a couple of years from now I might want to have another partner"*?

FG: Absolutely. For me in my bridge career, I've had about three or four serious regular partners, the most serious one being the one I had for about 15 years. We were very successful together and became very close friends, so it wasn't like I was really on the lookout for other partners. But throughout my travels in the bridge world, there were clearly people I liked and respected, and I'd think to myself, *If the time would ever come that I would need another partner, this is someone I'd definitely consider.*

HB: Is there a profile for you of the ideal partner? Presumably some-one who respects you and with whom you can get along, but in terms of bridge skills can you say, *"I'm looking for someone who can fit with my strengths and weaknesses"*? Can you characterize that profile from a bridge perspective? Or is it more psychological?

FG: For me it's more psychological. It's important for me that when I play the game I enjoy the person's company. It's too difficult to have to suffer the ups and downs of the game itself and have to deal with a jerk at the same time. So that, for me, is very important. In terms of the bridge style or bridge personality that the person has, I think one of my strengths as a player is that I'm fairly flexible and can pretty easily adapt to whatever bridge theories or bridge whims my partner happens to have. So I don't care so much about those details, as long as he is a highly-skilled player. That would naturally be important to me.

HB: Picking up on the "jerk issue" for a moment, I imagine that, this being a highly-competitive environment, there are an awful lot of people who are super-competitive and can be pretty unpleasant to be around. Or is it not that way? Do you get to a stage where, once you get to a high enough level, there's a mutual respect between competitors so that is taken out of the picture?

FG: Well, most of the very best players these days make their living from playing bridge. So it's important to keep up appearances. When you're playing, you're expected to behave yourself, to play profession- ally. Just as it's frowned upon for a professional golfer to throw his clubs at a tree or something, a professional bridge player is expected to behave himself at the table. So when people are playing, they are almost always well-mannered, but unfortunately, sometimes when they're not playing that doesn't happen.

I imagine that this can be traced back to the fact that there are only so many bridge players who can make a good living from the game, and there are more bridge players than there are slots for bridge players to make money. So, to some extent, there's back-biting and dog-eat-dog crap going on away from the actual bridge table; there are those who are trying to make themselves look better and make other people look bad. It can be kind of ugly sometimes, but the behaviour during the actual tournaments is actually quite good these days.

HB: You've spoken about making a living at the game. How does that work? What kind of money is out there? How do you get to that level? What's involved and what's the process?

FG: Well, first it should be said that most of the people who make their money from bridge are those who are actually doing the playing. In most tournaments, bridge is done on a team basis. A team will consist of two or three different partnerships, so four or six players. The very best players will play on a team for which there is a sponsor, who may or may not be a great player, who is paying everybody else to be on the team. Imagine that you are a very wealthy person and that you'd like to win a very big bridge tournament. You may be a decent player, but you're not good enough to win a national championship. But you could sponsor a team, and now all of a sudden you would have a chance to win.

HB: So if I were the sponsor, I would pay you appearance money and expenses to play on this team?

FG: Essentially, yes. And there might well be a bonus structure built in so that if we were particularly successful the players would get paid more. Nowadays in the United States there's probably somewhere in the neighbourhood of 20 or 30 major sponsors, each of whom will hire the best four or five bridge professionals they can to play in the biggest tournaments. The same things will go on in lower level tournaments and even club games. But in a club game, it's more of a lesson-type of thing, like when you're paying a tennis pro to play with you at the club.

HB: And teach you along the way.

FG: Exactly. So, that's how most bridge players make their living: by either playing and helping a lesser player learn, or to win a big trophy.

HB: Is there a difference between national championships or world championships and these sponsored events? Or is there a mixture between them? How does it work?

FG: It's pretty well all the same. As it happens, I just came back from a tournament that's designated as a national championship tournament. I think there were about 5,000 bridge players that went down to Dallas—that's where it was held—and there was a series of events that took place there. If you win the biggest of these events, you're considered a national champion. It's an open event—anyone can play—but many of the players who do play are either professionals or sponsors. It's the same thing at the world-level too actually. For example, the American team will often consist of five great players and one sponsor. Sometimes the sponsor is a great player; sometimes the sponsor is not a great player.

HB: I would think that in the long run, if the sponsor is not a great player, it would tend to bring down the results.

FG: That's true. It's much harder to win in that case.

HB: What does the notion of making a national team mean in light of the whole sponsorship thing? Is that related at all? Isn't that more of a Canadian or American bridge federation thing? How does that work?

FG: Each country does have its own national bridge federation, and one of the responsibilities of the national bridge federation is to select their team who will compete for the world championships. There are various ways that a team can be selected, but in the United States and Canada they basically have a single-event playoff to determine things. Anyone can enter the playoff; and it is often the case that a sponsor will enter the playoff and will win because the very best professional players want to get paid to do this.

HB: Let's get into a few more details of the atmosphere now. You go to a high-level event to compete; how often do you play? What's the stress involved, psychologically and physiologically? If you're completely unfamiliar with this world, you may think, *Oh, it's just a bunch of guys playing cards*. But there is considerable money involved, and there's a tremendous amount of mental processing that is going on all the time. So I can imagine that it would be really quite stressful both mentally and physically.

FG: It certainly does take a lot out of you. The tournament I just played in Dallas was 10 days long. You play about eight hours a day in two sessions. And when you're playing, you're just sitting there thinking and thinking, intensely concentrating. At the end of the day, it doesn't matter what kind of shape you're in, you're exhausted. By the end of the week, I've usually lost about 10 pounds.

HB: Really?

FG: Really. Now I don't know if that's typical, because for me it affects my eating and sleeping. I don't really eat very much during the entire tournament. That's one of the reasons why I lose weight.

HB: Sounds like a lot of fun.

FG: Well, it's fun in the sense that it's very satisfying when you win. Losing is terrible.

HB: Like anything, I suppose, at that level. And when you're playing these duplicate hands, I imagine that it's a very quiet environment; there's no shouting or screaming.

FG: It's very quiet, yes.

HB; Do you have a sense of how well you're doing as the tournament is going along?

FG: You find out the score at the end of every session. But a good player will typically know how he's doing well before that. Once a given deal is complete, a good bridge player will be able to tell what could have happened had everyone played perfectly. So you can judge yourself not only against perfection, but also against other tables that have the same hand. That's a skill that you just seem to pick up. It's kind of amazing that even not particularly great players, once the bridge game is over, can remember all the cards they had. It's not like they've ever tried to train themselves to do that; it's just by playing the game that your mind seems to somehow program itself to remember.

HB: And how long do you remember it for? Do you remember hands from two years ago, or five years ago?

FG: There are certain spectacular hands that I remember from 30 years ago.

HB: Wow.

FG: I'm not as good at this as I used to be. Getting older probably has something to do with it. When I was in my early 20s, you could ask me a week later and I could recite each of the 52 cards of every hand that I played. Nowadays, I only remember the vague features of the hand. But I remember the details of the spectacular hands, the really interesting ones.

HB: And at the time, when you were in your early 20s, did you do extra exercises for this, or any other mental training whatsoever?

FG: No. I actually have a clear memory of it happening: one day I could just do it and the day before I couldn't. It was amazing. It was as if overnight my brain had taught itself to do this.

HB: And how long had it taken you to get to that stage?

FG: I had been playing bridge for about two years at the time. It's funny: before that people had told me that one day I'd be able to do it and not to be bothered because I couldn't yet. And they were right: it just happened.

HB: It's just like a phase transition, as they say in physics.

FG: Yes, exactly.

HB: Although you say that you don't remember as well now as you used to; does that apply to conscious recollection as well? In other words, you might not just happen to remember every hand, but can you remember hands as well as you used to if you chose to?

FG: Most of the time. Sometimes I'll need a cue. If someone asks me about a particular hand that I don't recall, I might ask, "*Can you tell me what the spades were?*" And they'll respond, "*You had the ace and the king and the four of spades,*" and then I'll remember. That sort of thing will happen.

HB: Because memory plays such an important role in terms of real-time processing during the game, I might imagine that some players might use memory enhancement techniques. I mean, I'm a low-level bridge player; and one of the problems I have is that I tend to forget information while I play, and without that information you're in deep trouble. So if I wanted to get better, should I master some memory techniques, or is it simply a question of playing more and more?

FG: I think it's really just a question of playing more and more. I have no doubt that you could learn how to do it, you just have to play enough.

HB: Well, I guess.

FG: I promise you could do it.

Questions for Discussion:

1. In what ways does the partnership aspect of bridge make it unique among "mind sports"? Could you imagine a similar sort of competitive activity with teams of three people?

2. Is there a certain type of personality type that high-level competition in bridge naturally attracts? Are there similarities between those who are drawn to bridge and other activities?

3. To what extent do you think that memory development techniques associated with bridge could be applied to other areas? What, if anything, might they tell us about the way that memory works?

III. The Bridge Business

Educational software, online domination and special friends

HB: You've done a couple other things that I'm guessing other players at your level have not, namely you've done quite a bit of work in bridge education. You've taught people, you've mentored people, and you've also developed a very successful software company that offers not only bridge pedagogical teaching techniques, but also online bridge playing. I'd like to talk about each of those in turn, but first perhaps we can focus on the teaching. Does the act of teaching in any way help you clarify any bridge concepts at your much more advanced level? Or is all that so far beneath you that it's not even useful in that sense?

FG: No, I think it does to some extent. When a bridge expert has to teach a lesser player, it's very easy to take things for granted, things that for me I don't have to think about, things that I just understand. And by seeing the blank stares of the students who don't understand, and empathizing with them in realizing that I have to explain this in a way that is at a much lower level than I think about, it does give you some sort of insight into what the game is all about. By forcing yourself to put into words something that has always been easy and you don't have to think about, it helps you understand better, I think.

HB: I imagine it gives you some sort of satisfaction the way that teaching gives all sorts of people satisfaction: it's satisfying to interact with people, help them, mentor them. But I'm guessing there's another aspect of all this, which is furthering the cause that you believe in, getting more people involved in the game. Earlier in the conversation, you had darkly eluded to the fact that you're not too optimistic about the future, that we're a long way away from the heyday in the 1950s when everyone was playing bridge. Do you feel a sense of obligation to give back to the game, to encourage more people to play it? Or is it simply a case of enjoying teaching for the sake of teaching?

FG: I don't know if obligation is the right word. In some ways, particularly in the case of designing software, I realize that I have the ability to make a difference. Whatever the future of the game is I can make it a little bit brighter at least, maybe a lot brighter, and I care about that. It's not like I feel like I was put on this earth to change the world as far as bridge is concerned or anything, but I see that the work I do does make a difference. When it comes to teaching, maybe I make a difference there too. But with teaching I can only impact small numbers of people at a time, while some of the software I write has been used by millions. As we speak there are probably 15,000 people playing bridge on the site that I helped build, whereas when I teach, even if I taught rooms filled with people—which I've never done, actually—but even if I had, how many people could that affect? A hundred people, maybe? But the software I write can be used by the whole world.

HB: When did you start writing bridge-related software?

FG: I started experimenting with it in the late 1980s. At the time I had a "real" job and Sheri saw what I was doing with the software and said, *"You're doing wonderful things with this and we can make a difference with it. People are going to want to use your software."*

HB: So you began writing bridge software just because you thought it was a cool idea?

FG: Yes, exactly. At some point we had to decide how we were going to turn it into a commercial venture, to put food on the table and pay the mortgage and all that, but it was really driven by the love of the game, the software that we wrote. We started out by writing educational software, here's one of our CD-ROMs.

For the first 10 years or so we made products like this, and then we started turning our attention to the internet, roughly around the year 2000 when everybody started to have computers that were connected to the internet. We figured out ways to get people to play bridge over the internet and to learn bridge over the internet. And that has been our main focus ever since.

HB: Before we get to the online business, let's talk a bit about how the software works. Say I'm somebody who wants to learn bridge and I get one of these products, I get a CD version or an online version of your bridge educational software. What does it actually do for me? How does it teach me?

FG: Well, it's sort of like an interactive textbook. You read the rules the way you would if you were reading a textbook and then there are exercises at the end of each chapter, quizzes and things, to make sure that you understand the material. But because it's software as opposed to a book, we've been able to make it interactive. So in some exercises at the end of the chapter you actually play bridge against the computer and the computer tells you when you've made a mistake. For most people, that's very comforting because you can learn at your own pace: you don't have to worry about looking stupid in front of a class or have somebody yelling at you because you didn't get it right. It's a very relaxed way to learn.

HB: You're a bridge expert, but there's also the question of understanding pedagogy: what levels are the right ones to teach what to, and how exactly to go about it. Just because you yourself are an expert at something hardly means that you have some clear understanding of how to guide somebody through a learning process—in fact many people who are very good at something are terrible at teaching others how to do it, precisely because it's so natural to them. Was this something that you bounced off many people to get the right sense of things, or are you just instinctively insightful in terms of what you think are the best problems to go through?

FG: I think I'm instinctively not hopeless at it, but I definitely sought the opinions of players of various levels, repeatedly asking, *"Are we teaching this in the right order? This hand feels like it should be appropriate for someone who's been playing for two years, but what do you think?"* So we definitely asked lots of people and listened to what they had to say.

I think we did a pretty good job of it, but as you say it wasn't easy, particularly being an expert myself, trying to put myself in the mind of someone who had been playing for only six months or a year or two, understanding what they would be ready for. So we asked many players at those levels, and teachers and other bridge experts, and then made our best guess.

HB: You mentioned earlier that you were the first person of your peer group to have a programmable computer, which, if my memory serves, was a Commodore 64.

FG: Actually a Commodore PET.

HB: Okay. A Commodore PET. I guess what I'm wondering about is whether there's a connection between your interest in computer science and bridge. It doesn't surprise me that some body who was interested in programming, who is very logical and analytical and naturally had above-average memory skills, would be intrigued by, and be successful at, a game where one has to constantly do this high-level analysis, filling in who has what card at what time and how to solve these sorts of problems. It seems like an ideal fit, but is that just my sense of romanticizing things? Or do you think that there is some sort of connection?

FG: I think there's a connection between being interested in problem-solving, programming computers (which is really an exercise in problem-solving), and being interested in bridge, which is a different kind of problem-solving. But I don't know which is the chicken and which is the egg: if I was born with an innate ability and love for certain types of problem-solving, or if my exposure to computers when I was 10 or 11 years old is what brought it out in me. Perhaps everybody is hardwired to do this and you just need a stimulus that will get you thinking in that direction.

HB: I'd like to talk now about your business venture, Bridge Base and Bridge Base Online (BBO). Am I right in saying that BBO is one of the largest online bridge-playing sites?

FG: I believe that it's actually larger than all the other sites put together. We basically have a monopoly.

HB: Cool. And achieving that sort of business success takes rather different skills, I imagine, than some of the other things we've been talking about.

FG: Yes. And that is something that I would honestly claim to have no natural ability at all in. The idea that BBO has become a very success-ful business is interesting, because it was never intended at all to be a business. Basically, one day Sheri and I decided that there should be a free, high-quality online bridge site, just because that would be good for the game. At the time we were selling enough CD-ROMs that we could afford to put a little bit of time and money into the idea, so we made one. We had no idea at all that it would eventually become a successful business. But it grew and grew, and we got to the point that every day hundreds of thousands of people were logging in to our site. I think it's that case that if you have enough people involved in something, you can eventually find a way to turn it into something that makes money.

HB: So how do you monetize it if it's a free site? How does that work?

FG: The main source of revenue we have right now is that people can pay us one dollar to play in a mini-bridge tournament on the inter-net. If you're successful in the tournament, your ranking as a player goes up—your status in the world of bridge. Quite a lot of people are willing to pay one dollar to play to try to improve their ranking. Only about 10% of the people who log in to our site ever play these tournaments and give us a dollar, but there are so many people playing that 10% of them is quite a lot. It's very strange making money one dollar at a time, but if you have enough dollars it works.

When we started selling our CD-ROMs for $60, I remember thinking how unusual it was that it costs only a dollar to print one of these and we sold it for $60—it was almost like printing money. And $60 each adds up; and one dollar at a time adds up even more: you just need more people.

HB: Well, it costs only a dollar to print the CD-ROM, but there's the huge amount of unique investment to actually make it in the first place.

FG: Of course: once you develop the product it's pretty straightforward to make lots of them.

HB: One of the things people often hear about you is that you've taught bridge to Bill Gates and Warren Buffett. How did that come about?

FG: One of the many wonderful things that has happened to me as a result of bridge is that I've become friends with Bill Gates and Warren Buffett, largely through the software that I wrote. That's how I came to their attention.

HB: They started using it and they contacted you? Is that what happened?

FG: That's right. First was Warren. He just phoned me one day and said, "*I like your software and I'd like to order some more.*" And I gave him the sales pitch.

HB: You gave him the sales pitch? He already said he wanted to order some more.

FG: No, I mean I told him what I think he should buy and how much it would cost and everything. And when it came time to take his name and address, I realized who I was talking to, and I asked if it would be okay if I included a letter asking for some business advice. So I did that, and we began a written correspondence. As I said earlier, I don't exactly pride myself on being someone who understands business or has great instincts in that area, so I very much welcomed the opportunity to speak to someone who obviously knew what he was talking about.

Among other things he said was, "*I don't know very much about technology, but I may have some people I know who play bridge who do know about technology, and perhaps one of them could help you.*" Well, one of those people turned out to be Bill Gates. I eventually met them both in person: they were chartering a private train as part of a vacation, and they wanted to have a bridge leg of the trip. So they asked me to come and play bridge with them on the train.

HB: Who was the fourth?

FG: The fourth was a friend of mine named Sharon Osberg. I believe she knew Warren from being an executive at Wells Fargo, which was one of the companies Warren had an interest in. She had become close friends with Warren and through him she had met Bill. She's a world champion player, as well as a very impressive businessperson.

HB: An interesting foursome.

FG: Yes. An interesting foursome.

HB: And how long was the trip?

FG: Well, they flew me to Bozeman, Montana. And then I got on the train.

HB: This was the first time you actually met either one of them?

FG: Yes. The train trip was to Denver. It took two days and we played bridge the entire time.

HB: And the entire purpose of the trip was for you to play with them?

FG: Yes, exactly.

HB: So what was it like? Were they good players?

FG: Warren had been playing bridge for many years. I think when he was growing up his family were bridge players, so he was quite competent already. Bill was relatively new to the game at that point, but as you can probably imagine from what you've read about him, he has a voracious appetite for knowledge.

HB: So it probably didn't take very long to see some marked improvement.

FG: No, it didn't take long. They're both very competent players now. I have no doubt that if they'd dedicated anywhere near as much time to the game as I have, they would both be able to compete at the very highest levels. Of course, they both have very serious lives away from bridge, so that's naturally limited their bridge progress a bit, but they're both good enough to be able to compete very well. If they would play against people like Sharon and me, we're obviously big favourites to beat them, but they're certainly not going to embarrass themselves.

HB: That must've been a pretty cool experience.

FG: It was an amazing experience, and very refreshing. Here you have not only the two wealthiest men in the United States, but people whose accomplishments are historic in nature: these are people who will be remembered 100 years from now; they are obviously larger-than-life figures. And there we were, playing cards, and they were just two regular guys. That, to me, was very refreshing: that bridge was the great equalizer. At the time, I was just some kid from Canada; and here were these two enormously wealthy guys who have changed the world in so many ways, and we were all just laughing and having a good time playing a card game.

HB: And you kept seeing them after this?

FG: We did, yes. We have since played together in tournaments and socially.

HB: And they're both still passionate about the game?

FG: They are, yes. But since Bill retired from Microsoft and became so involved with his foundation, he doesn't play as much as he used to. Warren was never really interested in playing competitively; I think he enjoys the social aspects of the game. They're really the best of friends, the two of them. Because Warren doesn't really enjoy the tournaments so much, Bill will usually defer to him and play socially.

HB: Oh, they play as a partnership?

FG: No, they don't. Nowadays, when they play it tends to be in a social situation, as opposed to a tournament setting. Warren used to humour Bill more and be willing to go to tournaments, but not so much these days.

Life is strange. To think that I've had a chance to get to know people like that.

HB: Well, they've gotten a chance to get to know you as well.

FG: They probably think life is strange too.

Questions for Discussion:

1. Do you think that Fred's software and computer programming skills are somehow related to his natural bridge-playing ability?

2. Do you think that the success of online bridge will enable the game to survive, if not thrive?

IV. At the Summit

Conquering the world and then reverting to type

HB: I should have talked about this earlier, but I forgot. And you're such a charmingly modest fellow, it's not surprising that you didn't bring it up yourself, but it's probably worth mentioning that you have actually won a world championship in bridge.

FG: I have, yes.

HB: You'll be relieved to hear that I'm not going to do the typical sports-journalism thing and ask you, "*How did that feel?*"

FG. Thank you.

HB: But I'm going to ask a different question: *Did you look at this as a wonderful bonus to all of your efforts, or more as some sort of necessary culmination to complete your career?* You had mentioned earlier that you weren't a particularly competitive person. Had you been constantly saying to yourself, "*Gosh, if only I could win a world championship*"? Were those the sorts of thoughts going through your mind for all those years?

FG: Well, not at the beginning, for sure. I think that what happened was that, after I had been playing bridge for about 10 years, I began to get noticed by the better players in Canada, the elite players who represented Canada internationally, and they started to invite me to play bridge on their teams. At the time it was a massive honour: these people were my ultimate heroes. I wanted to be just like them and I was thrilled to have a chance to play with them. They were highly competitive people who were trying to be the best in the world, and if

I wanted to stick around them I had to force myself to become much more competitive than I ordinarily would have been, because they were expecting that.

HB: Otherwise you would have stuck out somehow.

FG: Right. Or I just wouldn't have been good enough. I never would've lived up to my potential. I really had to apply myself and say, "*Okay, I'm really going to do whatever it takes to compete at the highest levels.*" I guess, consciously or not, I underwent some kind of personality transition and became a competitive person. At that time I thought my goal was to become the best, and the best is to win the world championships. That was around 1990, so it took me 20 years to do it. But 20 years later, in 2010, I won a world championship. Several times before that I had come close.

HB: I was going to ask you about that.

FG: This is a gold medal. I have 3 silver medals and a bronze medal as well. So it took a while, but I eventually achieved my goal and became a world champion.

HB: Where did it happen?

FG: In Philadelphia—a bit funny, really. Nothing is wrong with Philadelphia, of course, but one of the things that bridge has done for me is, it has given me a chance to see the world. I've been basically everywhere: I've played bridge on every continent and in about 50 different countries. So it was wonderful to have finally won a world championship reasonably close to home, but given some of the exotic locations in which I've played bridge, Philadelphia felt a bit...

HB: I'm sure you didn't mind at the time.

FG: No. Not at all.

HB: So once you won the world championship, was your appetite whetted even further? Or did you revert back to non-competitive type? Did you say to yourself, *"Now that I've achieved my goal I can go back to being uncompetitive"*?

FG: Well, the final chapter has not been written yet, but I think that's the way it's likely to go. It's kind of strange, really. I still love the game very much, but I'm just not that interested in playing competitively right now. I can't say for sure that that's how I'm going to feel in a year or two, but for now I feel like I've done what I set out to do. I'll stay very interested in bridge, of course, but I'm not necessarily trying to be the best.

HB: And in terms of your development as a player, did you feel like in this 20-year period leading up to the world championship you were continually improving? Or was it just a question, when you were getting silver and bronze medals, of somehow not being lucky enough at the right time? Do you think you were a much better player in 2010 then you were in, say, 2000?

FG: I was certainly a much better player, in some areas of the game perhaps 100% better. In particular, the bidding in bridge is something you can become better at by experience: the more bridge hands that you see, the better your judgment becomes. Maybe it's a part of pattern recognition that you start to, consciously or not, compartmentalize things in your head; you think, I've seen enough other hands like this that I know what I'm supposed to do now. Whereas when I first started as a kid, although the raw brain power that I could process with was much more powerful than what it is now, I didn't have the experience to draw on and I was just making things up as I went along. So experience has value for sure, as does emotional maturity.

I think that as you grow up you become better at dealing with adversity. And in bridge, one of the most important factors that decides the outcome on any given day is whether you're going to play your best game or 80% or 40% of your best game. And the solution to that is mental focus: how well you can focus on the task at hand. If you allow yourself to get distracted by emotion, then you can't focus; it's impossible. The older you get, the better you get at shutting out other distractions like emotion or other things in your life. It's easier to do that when you're a more mature person.

HB: And this may compensate for the lack of, as you said, "raw brain power", or at least the diminishment of that raw brain power.

FG: Yes. As far as I can tell, most bridge players peak at around the age of 40, which I'm pretty sure is well beyond when they're at the peak of their mental powers.

HB: So it's a combination of mental processing and experience, processing power and emotional maturity.

FG: That's right. Of course, there are players who play well into their 70s and can compete, and sometimes even win, at the highest levels. There's also the occasional prodigy, a teenager or someone in his 20s, who may win a major tournament. It's a game where anyone can play exceptionally well on any given occasion, but for the average person, it likely doesn't get much better than when you are in your early 40s.

HB: Again, in terms of partnerships—because when you're winning these things, you're winning as a team—is there a sense that people are looking at matching age and experience, that if you're a prodigy at 18 you really should be with someone who's at least 40 or something? Does that play any part in the self-selection process of teams?

FG: You see some of that, but it will often be the case that the most talented people of a generation will naturally congregate together and form partnerships. Now, they would certainly benefit greatly, and do benefit, by speaking regularly to more experienced players, but they don't necessarily have to be their partners to do that; they can be their teammates, or they can just watch them play, or read what they have to say, or study their moves. One of the other cool things about internet bridge is that at any time, when you log on to our site, there are probably about 100 world-class players playing. You can watch their every move: you can see their cards and see what they do.

HB: That's a great learning tool.

FG: Yes. And if you're lucky, and respectful, the person might be willing to answer your questions too. You can ask them, "*Why did you play the 7 of diamonds?*" and they might well answer you. In addition, sometimes there will be commentators on the site who will do a sort of live, play-by-play analysis of the best players in the world, explaining in real time why we think he played the 7 of diamonds at that moment.

HB: Have you ever done that?

FG: I used to do that a lot, actually. But eventually my inherently anti-social nature came through and I decided that I didn't want to do it anymore.

HB: Well, that took a while.

FG: It did. But as we were building the site and trying to get more people to come, I was a draw because I was a very well-known player, so people would come to hear what I had to say. Now we're big enough that I don't have to be used as a draw anymore.

HB: So you can withdraw.

FG: Right. I can withdraw and be the natural recluse that I am.

Questions for Discussion:

1. Do you agree or disagree with Fred that the age of 40 is when most people are "well beyond the peak of their mental powers"? To what extent does this depend on how one categorizes "mental powers"?

2. Why do you think that Fred's desire to compete at the highest levels has abated considerably? How much of that was related to him winning a world championship?

V. Critical Times?

Demographics, gender and the future of bridge

HB: Let's talk about the future of the game now, something you're obviously passionate about. We talked earlier about these misconceptions that bridge is "just a game for older people". Clearly it's not, but that's the reputation. You said at the beginning of our conversation that you were not terribly optimistic about the future.

FG: Well, the expression that I sometimes invoke is that the game is about to fall off of a "demographic cliff". I believe that the average age of a bridge player in North America is over 70.

HB: The average age is over 70?

FG: The average age is over 70.

HB: Of course people are living a lot longer; over 70 is not what it used to be. But still, that's not what you want.

FG: No, it's not good. There are a few hopeful signs, as I understand it. Tens of millions of baby boomers are retiring, and one of the ways that bridge has been well-marketed in recent years is as a way to help keep your mind active.

HB: Right. Which, of course, it does.

FG: It does. It's not a scam or anything; it really does that. So some of the people who market bridge have had some success in getting baby boomers to take up the game, which might represent a shot in the arm of sorts. But maybe it just puts off the inevitable decline for another 10 or 20 years. It would be much better if college-age people learned to play bridge the way they did when our parents were in college, but that is something that I'm not particularly hopeful about, unfortunately.

HB: So let's talk about what can be done.

FG: Well, if there were some way to convince millions of people to try bridge, you'd get a lot of them sticking with it. The nature of the game is such that something like 10% of those who try it, will fall in love with it. It's simply a fabulous game: it just does that to people. So the way I think about it is that the marketing challenge lies in getting those millions of people to try it in the first place.

Now, that's difficult to do for a number of reasons. First, there's the image problem. If you're a young person, and if you've heard about bridge at all, you probably think it's a game for grandmothers. And even if you really love your grandmother...

HB: And I think most people do.

FG: Indeed they do. And most people have wonderful respect for their grandmother and might even think that if grandma thinks it's a wonderful game then it probably is a wonderful game, but they'll still likely conclude that it's not for them. There are a lot of alternatives these days for young people when it comes to entertainment: there are thousands of TV channels and a billion websites and apps on their phone. So bridge has a lot of competition, and that's a problem. Even if you can somehow get by this, "It's a game for old people" label, why is somebody going to want to try bridge rather than all these other things? It's a big marketing problem.

HB: Are there additional logistical issues involved given that it's a partnership game? So that, even if you think, *Okay, I'll give this a try*, you have to give it a try with someone; you have to find three other people if you are just starting out, and it's a bit of a barrier to walk into a bridge club.

FG: You make a good point, but I think there's a way around that using software. You can learn to play bridge by playing against robot players, so you don't really need a partner or other people to play with at first. In some sense that gives me hope, because that's going to have considerable appeal for some kid who likes technology and enjoys being able to play a challenging game on her phone against smart computer opponents. We have those sorts of programs already, the question now is just, *"How do we get ten million kids to try it?"* And I don't really know how to do that.

HB: Well, it sounds like partnering with school boards or something like that might work. I remember hearing about these Chess in Schools programs for students. Maybe you could do something similar with bridge?

FG: That's been tried, actually. I wasn't directly involved, but the way I understand it, both Bill Gates and Warren Buffett put up a lot of money towards this cause and coordinated a number of different, well-thought out approaches to school boards, describing all the advantages that a game like bridge would have on the development of the students. They offered to provide training material and bring in teachers and all that, and the program didn't get anywhere because of the layers of bureaucracy that they had to deal with. So it's kind of a sad story, actually. And the way I understand it, they had even selected school boards in advance that they thought might be most accommodating, but unfortunately they didn't get anywhere.

HB: One last thing. You talked before about the fourth on the Montana train—Sharon, was it?

FG: Yes. Sharon Osberg.

HB: Right. And earlier you mentioned your wife, Sheri Winestock. I'd like to ask about bridge and gender. One of the things that's long confused me about games like bridge (and chess) is why there's a distinction between male and female events and why so many of the best bridge players are men. Do you know why that's the case?

FG: Not for sure. Do you understand why most accomplished physicists are men?

HB: I have a theory, but that's not quite the same as understanding. But I have a theory. Do you want to hear my theory?

FG: Sure, because it's probably similar to my theory.

HB: Well, my theory is that statistically men are more prone to the level of highly obsessive behaviour that typically leads to maximum proficiency in these sorts of things, immersing themselves in activities like bridge or chess or physics to the virtual exclusion of anything else. I think it's also probably somehow linked to the fact that there's an established statistical correlation of males with things like autism and monomania. That's my theory.

FG: Okay. And do you think that some of this may be driven by pressures from society? That at least in past decades, bright young women have been expected to—

HB: Sure. I think there's a huge environmental component that reinforces this, but I suspect it's not solely attributable to that. But anyway, I'm not the one who's supposed to be answering the questions. I'm supposed to be asking the questions—which is much easier.

FG: All right, all right. Well, I think the same sort of explanation lends itself well to bridge, but I think there's something else too: the very fact that women's events exist, I think, might be part of the problem. Suppose that you are a brilliant young woman who happens to be talented in bridge, and you have a choice, say, in your early 20s to play in the men's tournament and maybe come in 7th or 8th or play in the women's tournament and maybe come in 1st and make a lot of money together with fame and glory. It's not that hard to be sympathetic to the woman choosing to play in the women's event, and continuing to do that.

HB: Sure. And these things continue to be reinforced.

FG: Exactly. But I think that the best way to improve in most activities, be they mental or physical, is to play against people who are better than you.

HB: Sure.

FG: And by playing in the women's events, they're setting a sort of bar for themselves which they can't go any higher than, because they can only play against the best women, and can't play against the best men. So it feeds on itself.

HB: And I'm guessing, based upon what you said earlier combined with my own experience, that once you have established that sort of "clubby" mentality, then the psychological pressure and the attitudes of people only get stronger.

FG: Absolutely. I mean, it's actually not uncommon to find an accomplished male bridge player who will categorically refuse to play on a team with a woman.

HB: Really?

FG: Yes. To me, that's a shockingly ignorant attitude. There are certainly women I know who play in the same league as someone like me. It is true that there aren't very many of them, that there are a lot more males than females at my level, but great women certainly exist. I don't understand the attitude some people have: that under no circumstances will they play with a female. But there are, sadly, people like that. And they don't outwardly seem like crazy, bigoted people, but they're there.

HB: So what percentage roughly, would you say would fit that categorization?

FG: I would guess that out of all of the top male bridge players, probably a quarter of them would have that sort of attitude.

HB: Wow. That many?

FG: Yes.

HB: Goodness. We talked before about getting more people interested; perhaps, under the circumstances, one way to do that might be to focus on younger female players.

FG: I think that would be smart.

HB: Indeed. Anything else you want to add?

FG: I can't think of anything. I think we covered a lot.

HB: Well, thanks for your time. It's been great fun.

Questions for Discussion:

*1. Are you surprised at the idea that Bill Gates **and** Warren Buffett were unable to convince school boards to implement an in-school bridge program? What, if anything, does this imply about the prospects for educational reforms in other areas?*

2. Why do you think that there are so many fewer top female bridge players (or chess players)? Is this, in fact, a problem?

The Joy of Mathematics

A conversation with Ian Stewart

Introduction

Counting Sheep

An engineer, a physicist and a mathematician are on a train that has just crossed into Scotland when outside of their window they see a black sheep standing in a field.

"How odd," says the engineer. *"All the sheep in Scotland are black."*

"No, no!" counters the physicist indignantly. *"You can't just generalize like that. All we can say is that only some Scottish sheep are black."*

"You idiots," sighs the mathematician, exasperated. *"Always jumping to conclusions. The only thing we can say with any certainty is that there is at least one sheep in Scotland which is black on one side."*

This joke might surprise you for two reasons.

For starters, you may be shocked to discover that there is such a thing as "mathematical humour" at all. It's not the sort of thing that most people are even aware of.

But more significantly, it might make you suspect that what mathematicians are—what they do, how they think and what motivates them—is very, very different from what any long buried tussles with long division might have naively led you to believe.

Of course, we frequently have a pretty superficial image of other professions. Policemen do more than chase bad guys, while doctors do more than dispense medicine. On the other hand, policemen do chase bad guys and doctors do dispense medicine. Mathematicians, on the other hand, typically spend no more time doing long division than the rest of us. And most aren't any better at it—or any more excited at the prospect of doing it—than anyone else.

So what do they do all day?

According to Ian Stewart, Professor of Mathematics at the University of Warwick and a highly acclaimed writer of both popular math books and science fiction, real mathematics is far removed from arithmetic.

> *"It's about form and structure and logical connections. If certain things happen, if a problem is set up in a particular way, it has certain ingredients. What does that tell you? How can you answer it? It's about problem solving, but it's also about seeing the kind of elegant structure that opens up a better understanding of whatever it is you're working on."*

If all that seems a little too abstract and highfalutin, jarring a bit too harshly with your own school day experiences, Professor Stewart will happily detail his mathematical motivations further.

> *"I like to work in areas, which, essentially, combine symmetry with the real world. Symmetries give you a built-in, beautiful, mathematical structure even before you start. It's got to be nice, because symmetries are always nice. And then I find some connection to the real world. Combining the two, that's where I feel happiest."*

Still too abstract?

Well, consider this. One of Stewart's most significant achievements came from applying mathematical symmetry to animal movement. And it happened, as so often tends to be the case in the world of basic research, quite by accident.

At the time he was an expert on the mathematical symmetry principles of linked oscillators (think of springs or pendulums that are connected in some way). But then one day a popular science magazine asked him to review a book on how engineers take inspiration from nature, which serendipitously drove him in quite another direction entirely—at least at first.

"There were chapters on vision and robots, but it had a chapter with a whole series of papers on animal movement. One of the things they talked about was symmetric gaits, a gait being a pattern. And as soon as I saw 'symmetric' I thought, 'Oh! That's interesting!'

"I looked at the patterns and they were just like our oscillator systems. So in my review, I put in a throw-away sentence saying, 'These patterns look like things I've seen in coupled oscillators. I wonder if anyone would want to fund that sort of research?' I think I said, 'Could anyone fund a robot cat?'

"About two days later I got a phone call from Jim Collins, who's a biomechanical engineer now at Boston. Jim was in Oxford at the time, and he phoned up and said, 'I can't fund a robot cat but I know some people who can. Can I come and talk to you?'

"He knew the biology, the physiology of animal movement, and I knew the mathematics of coupled oscillators. We talked to each other and we put together a series of papers explaining how we thought this connection went. And that's how it started.

"The beautiful patterns you see when an animal moves—think of a trotting horse, one diagonal pair of legs hits the ground and then the other diagonal pair—it's absolutely perfect rhythm and it's this beautifully symmetric movement. That's one of the standard patterns you expect to see in a network of nerve cells, which is presumably what is controlling the movement."

So now, suddenly, we've gone from networks of coupled oscillators to neurophysiology.

And that sort of leap, that rapid interplay between mathematical structure and the world around us, happens all the time, every moment.

"Math is in everything we use, everything we do. If you're aware of this, from the moment you get up in the morning you start to see it. I look at the clock: it's one of these modern clocks that sets its own time by radio. Without radio, that clock would not work in that way. Well, radio works because of a whole pile of mathematics. The

engineers have to know the math. We don't need to know it to read the clock, but in order to manufacture the clock somebody there has to know a certain amount about that. And that's just one example.

"It's absolutely everywhere. I often feel that we should have a little red label that says 'Math Inside'. And if we did, just about everything in the world would have these labels on them.

"I think people underestimate how much math is actually required, and how much our society consumes. In any given part of it, you personally may not be using this most of the time, and it's not the math you did at school.

"School math is about one-hundredth of a percent of the mathematical enterprise. It's not the tip of the iceberg: it's an inch or so on the top of the iceberg. People simply don't understand just how much math is going on. And nearly all of it—maybe not straight away, but eventually nearly all of it—starts to become useful somewhere."

Except, perhaps, for one-sided Scottish sheep.

The Conversation

I. Fear and Loathing

Mathematics and the wider world

HB: One of the things that's happened to me a lot, and I'm guessing it's happened to you an awful lot more, is that I'll go to some party, and I'll talk to some very educated, interesting person who's a historian, or an archaeologist or perhaps even a lawyer—although they tend not to be nearly as interesting. We'll be chatting and when the conversation turns to our backgrounds and I mention anything to do with mathematics, he'll get this look in his eyes, almost like a deer caught in the headlights, and say, "*Oh, I don't know very much about that: I'm not one of those types of people.*"

Does that happen to you a lot?

IS: It does. Perhaps not quite as much as it used to, but it happened about a month ago while I was on holiday. We were in the Caribbean, in Barbados, and we were at the manager's cocktail party in the hotel. I was talking to this very articulate, intelligent couple—I think he was something in the city—and there was the usual, "*What do you do?*" And I said, "*Well, I write books.*" I've learned not to say that I'm a mathematician straight out, you see.

HB: You ease into it gently.

IS: That's right.

And he asked, "*What sort of books?*"

And I replied, "*Well, popular science books.*"

"*Oh,*" said the lady, "*What's that?*"

"Well, for example, I've done one recently on the great mathematical equations and how they've influenced the course of human history and culture."

And she looked at me and said, *"I'm afraid that's way over my head."*

And I thought, apart from the word "equation" I haven't actually said anything yet.

HB: There's this defensive reaction immediately.

IS: Yes, there's this defensive reaction. So we moved on to a different topic and that was fine. But I thought, *It is actually possible to converse at that level without that kind of reaction, even if you know absolutely nothing.* I mean, I didn't know anything about what her husband did. But I didn't start off by saying, *"Oh, banking! No, I don't understand that."*

HB: *"I can't even have that conversation!"*

IS: Exactly. But I do get the impression that it's not quite as common as it used to be. It used to be almost de rigueur that people would respond in this way: *"Oh, I was never was any good at math when I was at school."*

I actually think it's a defense mechanism.

HB: What's it caused by? I mean, presumably it dates back to experiences at school, when they felt terrorized by some mathematics teacher.

IS: I think, particularly in the past, mathematics at school was actually a rather daunting experience. If it wasn't something that you were naturally good at, then it was tough. And the really tough thing about it was that if the answer was wrong, it was wrong.

The teacher asks a question, you say the answer is 17, but it's actually 16. You can't say, *"Well, that was close!"* Maybe nowadays you can, but you couldn't in those days. They would just say, "*Wrong.*" There was no wiggle room, no way out.

I think the other problem—which in some ways was greater in the past, then it eased off a bit, but now it's become a problem again—is that there is so much emphasis on the nuts and bolts of doing the maths, that nobody gets told why they're doing it, what it's for, what role it's played in human history, all of these things; because there isn't room for that in the syllabus. There isn't room in the school's timetable to go into that kind of thing. The poor teacher has got enough trouble going through the prescribed material and ticking the boxes to say that this is what the kids have done. In Britain these days, at least, essentially everything the teacher teaches in a week is prescribed, and they have to certify that the class has done certain things.

Of course, the bureaucrats think that if the class has done it and the boxes have been ticked, then they know it. And then we move on to next week. The idea that what we do next week depends on what we did this week—that kind of gets lost. And the idea that even if they've been through the material, maybe they don't really understand it—that too gets lost.

The real problem with mathematics is the way it builds, step by step, on everything you've done before. It's very hard to jump in afresh at some higher level. Once you're lost, you stay lost.

I always found with my students, that when they came in and said, "*I don't understand this or that,*" of some lecture I just gave, I would look carefully before realizing, "*Ah, that's actually not what you don't understand. What you don't understand is something from three lectures ago. Which is what we're now using. This thing would make sense if you had fully understood what you were doing three lectures ago.*"

HB: But these are mathematics students. I want to get back to this person at the party you were speaking of just a moment ago.

It seems to me that there is a sort of dichotomy going on when we are talking about people not understanding things. We all have areas of expertise, we all feel more comfortable talking about some things rather than other things. A few moments ago you were talking to me about Egyptology and your interests there. I don't pretend to have any experience whatsoever in that but I'm happy to have that conversation.

But when one talks about mathematics or the hard sciences, and particularly mathematics, there is this reaction: you can almost see the fear and the terror in people's minds. And I think to myself, on some sociological level, that that's very, very different than anything else. If I were to say, "*I study ancient Babylonian history,*" or "*I study the linguistic dialects of the Icelandic people,*" most people would reply with something like, "*Oh, that's very interesting.*" They may not be interested in it at all, but they wouldn't be running for the hills either.

IS: I suspect that if you could wire them up you would discover serious physiological signs of terror. In fact, I did once suggest that we should be asking for danger money to do mathematics because we're doing something that everybody else is scared of.

HB: But that's a real sociological problem at some level.

IS: It is, and they've acquired this feeling of fear from somewhere. I think some of it is this unrelenting nature of the subject. You feel you ought to be able to handle it. It can't be that hard, can it? And yet, there is a kind of mental block. Once you've got the mental block—without wishing to be nasty to teachers who have a very difficult job to do—it's so much easier to put somebody off a subject with a bit of bad teaching than it is to maintain their interest with good teaching.

HB: Do you think that anybody could enjoy mathematics? Or do you think that certain people are precluded by their neurophysiological makeup? You'll hear people say, "*I'm not a math person. I just can't do it, I'm not hardwired in some particular way to do math.*" Is that actually something that holds water with you?

IS: I don't think there's a kind of hard and fast barrier, but I think some people are naturally more able to handle mathematical thinking than others. I think you can train people to do better, but you can also probably train them to do worse—that may, in fact, be easier.

Young children mostly seem to be very interested in numbers and shapes and things like that. And at some point they start to just switch off and lose interest. I really don't know if that's because their schools have somehow trained them not to be interested in some way, or whether that's a fairly natural thing for most kids to do.

One of my problems with all this is that I didn't really experience that problem. I found maths interesting. But if you're a kind of people person, if you enjoy games and sports and communal activities and all that sort of thing, then mathematics is a very abstract, austere subject. It's just these number things, and you do these sort of incomprehensible things with them.

I do think that teachers—with some reason—tend to tell you how to do things but not why you do it that way.

HB: You're very careful and you're very generous to teachers. I tend to be a bit less careful and perhaps less generous.

Two of the things that you mention specifically in your book *Letters to a Young Mathematician*, in terms of the delusions and misconceptions that many people have from their school days with mathematics, are, firstly, the equivalence of mathematics and arithmetic—this sense that that's all mathematics is: calculating, adding and subtracting, long division, what have you. And secondly, this notion that everything is at the back of the book: that the field is a static, self-contained whole with no room for discovery, creativity, and ingenuity.

When I've told these self-proclaimed non-math people, that "*No, mathematics is about beauty, elegance, discovery, creativity.*" They look at me as if I'm from Mars because that is so completely different from their experiences.

Is it more a question that they haven't been exposed to what mathematics is and what it can do? Or is it more a question of their own lack of abilities?

IS: I think it's mostly that they have not been exposed to this. I was defending the teachers because the teachers, even if they want to do that, are often not allowed to. But it is also true, in a sense proving that mathematics must be more than what most people think it is, that most mathematicians can get very, very good, well-paid jobs. It's one of the best-paid professions, in terms of degree subjects. You earn more money as a mathematician than almost any other degree subjects.

HB: That's why you went into it, presumably.

IS: Well, of course. But this is a big secret that people don't seem to understand. Because there is such a demand for trained mathematicians in so many different kinds of jobs, very few go into teaching.

Which means that a lot of the people teaching mathematics weren't really trained to a high level in the subject. If you're trying to teach something, it feels very uncomfortable if your understanding is only three pages ahead of the class. I wonder if the teachers themselves are scared because someone's going to ask them a question and they won't know what the answer is.

HB: They may be scared, but they certainly won't have the appreciation to be able to inspire the student, to be able to give her perspective, to be able to portray a bigger picture of what the whole point is of what is being done in class.

IS: That's right. I was very fortunate. I had a couple of really good math teachers from about the time I was 13-14 onwards. They went out of their way to show those of us who were interested in and good at the subject that there was much more to math than you would have thought just going through the standard school stuff.

I never quite understood why people think that because the answers are at the back of the book that means that the book contains everything. It's just the answers to those questions. But I suspect it's because they'd actually like to be told that the book really does contain everything because if not, there's even more of this stuff that they're going to have to do.

I vividly remember Dame Kathleen Ollerenshaw, who's a very well known British mathematician and educator (she was also the Mayor of Manchester for a time), telling the following anecdote. When she was still at school, she said to one of her friends that she wanted to be a research mathematician. When the friend asked why, she replied, "*Well, I want to create more new mathematics.*" And the friend looked at her and said, "*Isn't there enough of it already? Please don't make more for me to learn!*"

If you're not enjoying it, if somehow your interest has switched off it's very hard to get it back. And as the subject moves on, you're left behind. It rapidly comes to the point where you just shut up shop and say, "*I don't want to face this. I don't want to do this.*"

But we insist on dragging the students through the math courses anyway. And I'm not sure there's an answer to that, in the sense that if you said, "*OK, then you don't have to do it,*" I don't think that would work either. Because if you say that, you're in effect saying, "*Here's a whole pile of jobs you'll never be able to do,*" and it's not just the standard list.

HB: And you won't be exposed to these ideas...

IS: That's right. There will be all sorts of times in your life when you will not understand what's going on. And if you'd stuck with this subject you maybe would have understood it, or some of it. Although we don't use school math every day in obvious ways, it informs our background thinking on a lot of things.

It's much easier to think about a newspaper report on some new medical breakthrough or whatever, which will quote a few statistics. And if you can just look at those figures and not be put off by the numbers, then you're already halfway towards understanding what the article is telling you about. It may be rubbish, and often is: it depends on the newspaper. But if you understand the statistics you have much more chance of judging this. I think that as an informed citizen there is a certain amount of basic math that you need to know.

It's a pity that we call it mathematics at school because the real subject for most of us is arithmetic. That's what sticks in people's minds. They think mathematicians do very big sums: research mathematicians just do bigger sums and invent new numbers.

Although when the wife of one of my topologist friends at Warwick expressed this particular view of it all being just big sums, he said instinctively, "*No, it isn't!*" before pausing for a moment and admitting, "*Well, actually, at the moment that's exactly what I'm doing.*"

HB: OK, well let's strike that one from the record.

I do want to get back to what mathematicians actually do and what mathematics really is so we can address some of those questions.

But before I do, I want to get back to the importance of a good teacher.

You said that you had very good teachers growing up, but in my experience that's a fairly common reaction. If you ask virtually anyone who has a career in the mathematical sciences the question, "*Did you have a particularly formative and influential experience with a high school teacher?*" They will say that yes, they were fortunate to have had a great teacher in math or physics or chemistry or what have you, that led them towards a deeper understanding and greater inspiration.

IS: Absolutely. From time to time we have conversations like this in the common room of the department with other faculty members, and yes, they all say that. It seems to be a very common experience. Also this: if you get enough exposure to real mathematics or something interesting from one good teacher, I think there's also some sort of threshold achieved, where once you're past the threshold, even if you then got a bad teacher or a less inspiring one, you're now immune. You know already. I went to the local library. I took out books—I started reading math books from the library (as all teenagers do, of course). But it didn't matter after that what the teachers did. I'd figured it out. I knew there was much more to it. I knew you could do creative work.

HB: There's another point that you mentioned in *Letters to a Young Mathematician* that I think is worth highlighting, and that is the defense against what I would call false romanticism.

Occasionally one hears comments like, "*Oh, all this mathematics, all this science, that's a poor way of looking at the world. I'm a romantic: I believe in beauty, elegance, and so forth. And you with your numbers and ideas, you somehow cast a pall on the true beauty of the way the world is.*"

You use a very specific example to confront this attitude when you talk about a rainbow. You say that when you look at a rainbow, you're not only able to appreciate the beautiful colours and the way they look but you're also able to appreciate why you see the same depiction of a rainbow as someone else does, and that's because of the symmetry of the raindrops and where the light comes out.

You're able to trace a definite path to symmetry, to a geometrical property, which gives you a deeper understanding of why we all see rainbows when we look in different positions. And for any so-called romantic who would claim, "*Well, that's all mathematical mumbo-jumbo, that somehow takes away from the beauty and the spiritual moment of looking at this rainbow,*" you would reply, "*No, this in fact, enhances the beauty for me, it enhances my understanding. Because it adds my mathematical awareness and my geometric sensitivities to the aesthetic sense of the rainbow.*"

IS: Yes. I do not buy this argument that if you understand something, you somehow destroy the magic. I just don't think that's true: we are not that sort of animal. We evolved over millions and millions of years in the natural world. We have an appreciation of the natural world in our heads for all sorts of reasons. Where our sense of beauty comes from is not clear to me. It's not obvious that it has survival value and I can't tell you an evolutionary just-so story about it. But it's in there and it's in all of us.

Now some people, certainly with training, may appreciate the finer points of a painting or a sculpture or something in a way that I can't. But I don't think it makes any sense to point to a rainbow and say, "*I am such a wonderful person and I know nothing about the science behind that rainbow. Therefore it's a much more valid experience for me than for you, who, stupidly, has actually figured out why it happens.*"

No, I think knowing more about it improves the experience because it puts it into a context. I can look at a rainbow and say, "*Now, I actually know something of the mathematics behind that. And that same mathematics comes up in various other places. And let me show you some of those. Let me tell you about related things that you might not realize are related.*"

So it enriches the experience.

I can't prove I'm right. I can't prove that I get just as much pleasure out of a rainbow as somebody who doesn't know how it forms. But, that's because nobody really knows how to do that experiment. But I don't think there's any serious evidence that being aware of the causes of things somehow spoils the effect.

It might if it's a piece of stage magic. If I don't know how the magician saws the lady in half, I'm feeling surprised and shocked as he does it and think, "*Wow! That's clever.*" But if I know how it's done, I tell myself, "*OK, that's clever but I know how to do that.*"

My wife and I once had lunch with a television magician, Paul Daniels, before a broadcast. He was showing us card tricks—he always keeps in practice, he wasn't trying to show off or anything, he was just using this as an opportunity to practice his tricks. It was simply astonishing what was happening to these cards in front of our eyes. It was right in front of us and yet you could not see how—I mean, it changed into a totally different pack at one point: it started as red backs and then ended up as blue backs. The entire pack! I don't know how that happened.

If he had shown us how to do it, it might have spoilt that. But a rainbow is not a conjuring trick: it's a natural phenomenon. And the mathematics is extraordinarily beautiful.

Questions for Discussion:

1. What first comes to mind when you hear the word "mathematics"?

2. Why do you think so many people are "math-phobic"? Is it possible to change that?

II. Doing Mathematics

An insider's view

HB: So, let's get back to what mathematics really *is*. We've talked about what it's not, what people might be under the impression that it is. They might think it's equivalent to arithmetic, they might think it's just a static enterprise of calculating bigger and bigger numbers, they might think that what mathematicians do all day, perhaps with the assistance of computers, is to do really, really complicated long division questions or something like that, because that's what they've experienced themselves.

But what is mathematics, really? What is it that you're doing when you're doing front-line mathematical research?

IS: I think mathematics is about form and structure and logical connections. If certain things happen, if a problem is set up in a particular way it has certain ingredients: what does that tell you, how can you answer it? It's about problem solving, but it's also about seeing the kind of elegant structure–you hope it's elegant because good mathematics usually is elegant—that opens up a better understanding of whatever it is you're working on.

There is this broad-brush division of mathematics into pure and applied. You can argue about it but it actually does capture a certain distinction in the broad way that various people think about whether one is driven by real-world problems or the inner beauty of mathematics. But you can be driven by both. And to some extent, I am.

HB: And many other people were, as you pointed out earlier. Gauss, for example.

IS: Yes, this is a fairly modern distinction between pure and applied. You can do both. You can wander from one to the other—a lot of the most interesting stuff is in this no man's land where they overlap—increasingly so, in fact. Actually, in contemporary pure or applied math departments—there are still a lot of places that separate them—there's much more understanding that we're all in this together: *You guys like to do it one way, we like to do it another way, but if we put it together it's much more powerful.*

I like to work in areas, which, essentially, combine symmetry with the real world. Symmetries give you a built-in, beautiful, mathematical structure even before you start. It's got to be nice because symmetries are always nice. And then I find some connection to the real world. Combining the two, that's where I feel happiest.

But if you would observe me at work, and watch what I'm doing, a lot of the time I'll be working on the pure mathematical end: trying to figure out what this structure is telling me. *Why does it work like this? Is this the right model for the problem in the first place? How do we understand this? What does it do?*

So I will be mostly thinking as a pure mathematician about the mathematical problem, but every so often, especially when I get stuck, my collaborators and I (because I tend to work with small groups of people) will take another look at the application, take another look at the real world. My main mathematical collaborator is very good at going off and talking to experimentalists or people who understand the applications better than we do, and asking them questions. You can't get that kind of thing by just reading the research papers because they don't tell you what you want to know.

When you're starting to put the math and the applied problem together, the kinds of questions you ask may not be the ones that people are used to asking in that subject area.

HB: In fact, the really big breakthroughs happen when they're not.

IS: That's right.

HB: One of your many, very interesting results is this idea of linking mathematical oscillations to animal gaits. That's something that I'm sure most people who are neither scientists nor mathematically inclined would find fascinating. They may think, "*My goodness! This is what a mathematician does, he does something like this.*"

Tell us the story about how that came about, because it's a fascinating story. Not only because of what you did but also how it happened, what the process of discovery was like.

IS: Well, it's an example of the indirect way in which really interesting research ideas arise. My friend Martin Golubitsky, an American mathematician I've long been collaborating with, and I were at a conference in Northern California. We'd been doing work on the mathematics of coupled oscillators: that is, systems that can vibrate periodically and you hook them together in some sort of network.

We'd worked out some of the properties of this in the car on the way back from the conference with nothing to write on. There were four of us in a Mini: three mathematicians and a physicist.

HB: So you were forced to collaborate.

IS: Yes. Martin and I were stuck in the back. He had picked up a crate of wine from his favourite vineyard on the way and we were heading for the airport hotel. We sat there in the back of the car and we decided to work out in our heads the basic mathematics of a ring of coupled oscillators. We eventually wrote this up and published it.

Then some time later I was reviewing a book, which was about engineers taking inspiration from nature. It had chapters on vision and robot vision but it also had a chapter with a whole series of papers on animal movement. It was robot engineers who wanted to make robots with legs for various purposes, and it was about what the biologists and zoologists had found out about the mechanics of animal movement.

One of the things they talked about was symmetric gaits, a gait being a pattern. And as soon as I saw "symmetric", I thought, *Oh! That's interesting.* I looked at the patterns and they were just like our oscillator systems. So in the review, which was for *New Scientist* magazine, I put in a throw-away sentence saying, "*These patterns look like things I've seen in coupled oscillators. I wonder if anyone would want to fund that sort of research?*" I think I said, "*Could anyone fund a robot cat?*"

About two days later I got a phone call from Jim Collins, who's a biomechanical engineer now at Boston. Jim was in Oxford at the time, and he phoned me and said, "*I can't fund a robot cat but I know some people who can. Can I come and talk to you?*" He knew the biology, the physiology of animal movement, and I knew the mathematics of coupled oscillators. We talked to each other and we put together a series of papers explaining how we thought this connection went. And then various other colleagues got in on the act. From time to time, every so often, there's another paper published on this idea.

The beautiful patterns you see when an animal moves—think of a trotting horse: one diagonal pair of legs hits the ground and then the other diagonal pair—it's an absolutely perfect rhythm and it's this beautifully symmetric movement. That's one of the standard patterns you expect to see in a network of nerve cells, which is presumably what is controlling the movement.

This was a great revelation and it came about by a series of rather strange accidents. If you involve yourself in lots of different things—I'm a bit of a butterfly brain, I'm interested in many different things—from time to time one of them will strike sparks off another one. I think you're creating opportunities for that kind of connection, which you would never notice if you were stuck in your own narrow, little field.

HB: I think "eclectic" is a better term than "butterfly brain".

IS: I think butterfly brain is more accurate, actually.

I sometimes maintain that there is virtually nothing that I have taken an interest in at some point in my life that hasn't turned out to be useful in some way, somewhere else. All sorts of courses that I took as an undergraduate that I thought would be irrelevant to my research, every so often something from them pops up, and I say,

"Oh, hang on—I know how to do that. I did my Galois theory course as a student at Cambridge and for this little piece of the problem, that's exactly the tool I need." You've got this kind of toolkit.

I said "butterfly brain" because my knowledge of this stuff is often not very deep. There are deep thinkers who seem to know not just a lot of areas of math, but they know a lot more about each area than I do. They've clearly spent a lot of time really understanding it. I know a few areas fairly deeply and an awful lot on the surface about other bits and pieces.

HB: But it takes all kinds.

IS: That's right. The mistake is to think that there's only one way of doing these things.

HB: One of the things I really wanted to highlight here is to elucidate what is exactly going on in the mathematical process.

Again, Stewart's not doing some long division question with million-digit numerals. What he's doing is seeing patterns, seeing connections, seeing forms. You're looking at symmetry—symmetry has played an enormously large role in your research, in your framework, from your PhD thesis—or perhaps even well before your PhD thesis—onwards.

IS: Yes. Looking back I can see that there was a common theme that runs through the sorts of mathematics that interested me early on. It was the algebra of symmetry as well as the geometry: it's this very interesting, very powerful combination. Clearly that appealed to me. I was lucky, I think, to find just the right area that really suited me. But if you keep looking around, eventually you'll find what fits. I got into the area and thought: *I really like this. This is really working. This is better than what I was working on.*

Because what I started working on was fairly closely related but rather more algebraic—rather less of a connection, certainly in the way I did it, with the real world. But after about five or six years of working in that area, I started to get the feeling that actually I ought to be doing something else. If you look at the published papers you can see that I was trying different things: dipping my toe in here, dipping my toe in there. And then suddenly I found what's basically the area I'm still working in.

Thirty years, then, in roughly the same area, broadly defined— although it's moved, of course. Actually, I'm now much more interested in the dynamics of networks that may or may not be symmetric. We have a much broader notion of symmetry, if you like. We're still looking for structure, we're still using a certain amount of the same mathematical viewpoints, but there's a whole pile of new stuff which as far as we can see you just have to invent new methods for. There isn't anything you can just take off the peg for these problems.

HB: I'd like to spend a bit more time on the process of discovery and what's going on in your mind when you're working on a problem, chewing on a problem, thinking night and day about a problem.

There was this specific anecdote that you talked about in *Letters to a Young Mathematician* when you were doing these doodles, with colour coding, of nodes of a network. You were stuck on a problem and you were trying to find a way out to get unstuck.

IS: Yes. Marty Golubitsky's postdoc had come up with an interesting network of oscillators that did something that symmetric networks ought to do. He could prove it but it wasn't a symmetric network. It was close but it wasn't the same. We were racking our brains to work out why this asymmetric network was acting as if it was a symmetric network, and getting pretty much nowhere with it.

Then a friend of mine invited me to a conference in Warsaw. And another friend that was in Krakow decided that we should go down there for the weekend. So I was sitting on a train going from Warsaw to Krakow and I started drawing little pictures of this network. And then I started colouring in the dots. Some of them were synchronized with others and so they were doing exactly the same things, so I coloured them all the same colour. And I ended up with this picture of 4 different colours of dots.

HB: And you were doing this just because you were playing around?

IS: I was trying to pass the time. I wanted something to do. It was a rather boring train journey, to be honest. Then I looked at the picture and said to myself, *Hang on: every green dot has an input from every red dot. Every red dot has one from a blue dot. If I draw the picture, not of the whole network but how the colours are connected to each other...look, it's symmetric.*

By the time I got home I pretty much had the whole thing worked out. I'd realized how to formulate the problem. I'd realized why this behaviour that our postdoc had observed was actually happening. Not only could I prove it, I could explain it. It had become obvious. It had opened up a whole new line. We're still working on it. We're currently trying to apply it to visual illusions and a phenomenon called "binocular rivalry": you show one picture to the left eye, one picture to the right eye, and what the brain perceives is neither of those pictures, or one picture alternating with the other, or a picture that's some sort of funny mixture, or a mixture of pictures alternating with each other. It's quite strange.

HB: So, the neurons are processing things differently or...

IS: Yes, the simplest one is that I show your left eye a vertical grid of black and green lines and the right eye a horizontal grid of black and pink lines, and then what people will see may be the green grid alternating with the pink grid. Or they might see a pink and vertical line combination, which is not what either eye is seeing: it's got the colour from one eye and the direction from the other.

HB: So different people will see different things?

IS: Different people will see different things but a given person will see much the same thing in any given experiment. There are some simple phenomenological models: they're not the exact brain wiring involved but they're some sort of a map of the kind of brain wiring involved, probably. If you design a simple network to try and recognize patterns, you can come up with precisely this kind of behaviour in these model networks. Our symmetry mathematics comes into this, our network mathematics comes into this, and it actually looks like you can make some predictions and do some experiments.

HB: That's what I was just going to ask: so it does lead to some type of experimental prediction?

IS: Oh, yes! We haven't done these experiments yet but we know people who maybe will do them for us. But in principle you can do it. There is a definite prediction, there is a definite experiment, and it will either turn out to be correct or not. And if it's not, maybe we can modify things. If it's correct, then we try to move on to another one. But it seems to make a lot of sense mathematically and the neurophysiology makes sense. These are very simple networks that could have easily evolved.

HB: Presumably, then, another whole link is to be starting from the mathematics and say, *"If the neurons are processing information in this particular way, according to these symmetry groups or principles or what have you, then what are the neurophysiological conditions in order for that to happen? And what would it be like if we imagined this was necessarily the case, and we'd scale this up neurophysiologically? What would happen on a larger scale?"* It might give us a deeper understanding as to how the brain actually works.

IS: You ought to join our group. Well, that's right. I think that every so often in a research career something happens—it may be very simple, it may not be a great insight in its own right—but something in your brain says, *"Hey! That might really work. That's what's going on here."* Or at least, *"I think that's what's going on here. And I can do something with it. I can make some serious predictions, and we can go and take a look and see if they hold up."*

　　If they hold up, develop it further: if they don't, back to the drawing board. Don't give up on a good idea just because it doesn't work.

But don't persist for too long if all the experiments tell you, "*No, it's not like that. You're just crazy, it's wrong.*"

We've derived it from a lot of experiments and it's consistent with pretty much everything we've seen. But it is possible to make new predictions. In fact it's actually better than the animal movement from that point of view.

HB: Really?

IS: Yes. We can make a lot more predictions from this idea than we can from the animal movement idea.

HB: As you were talking before about animal gaits, it made me think of something that I'd once heard. My understanding is that people actually got animal movement quite wrong. If you look at some 18th-century paintings of horses, they have all their legs extended at the same time going over fences and so forth. Because that's the way it appeared to them in their snapshot but they weren't actually depicting their movement correctly. Is that correct?

IS: Yes, that's right. If you go into what looks like a nice, real English pub—oak beams, black and white, looks like it's been around since the 14th century or something like that—and you look on the walls, you will often see these hunting prints. And, yes: they will have horses going over a jump and the legs are splayed out. Nowadays, we are used to watching movies and photographs, so we know what it really looks like; and it doesn't look like that.

But it sure looked like that to them. And there's a whole history there. One of the big problems from Aristotle right through to some-where near the end of the 19th century was, *If a horse is trotting, are there times when it's completely off the ground?*

Leland Stanford Jr., who founded Stanford University, supposedly had this bet with somebody for $25,000—which was a huge amount of money at the time—that the horse really was completely off the ground while running. Eadweard Muybridge invented, among other things, a camera that took very rapid photographs because it had a very quick shutter. He lined up about 25 of these cameras in a row with trip wires and they trotted a horse past: click–click–click–click. They took all these photographs, and one in the middle had the horse completely off the ground.

HB: So he got his money?

IS: Yes. Well, so the story goes.

HB: Maybe that's how Stanford was founded.

IS: Could be. Muybridge produced beautiful books of animals and people moving in this way. Eventually this became the movie industry, this was the basis of movie technology. They started trying to automate this: "*Can we do it with one camera and not 25?*" Well, what's a movie? It's a whole series of stills one after the other.

But until not much more than 100 years ago, people had no idea how animals actually moved. Muybridge's work was so striking and attracted so much attention because suddenly you could freeze an animal moving, and you could see what it was doing. It must have been amazing at the time. Imagine the first time ever you could see this kind of thing.

Questions for Discussion:

1. What role do aesthetic factors and inclinations play in mathematical research?

2. To what extent is the distinction between "pure" and "applied" a meaningful one in contemporary mathematics? Are there examples of some types of mathematics that are indisputably "pure" or "applied"?

III. Teaching Mathematics

How to get unstuck and other valuable lessons

HB: We've talked a little bit about what mathematics is: what you've done, how it felt, what you've accomplished, what it really means to be doing mathematics. At the beginning, we spoke of the frustrations that some people have with mathematics.

Now I'd like to move to something which ties into another very strong aspect of your particular history and proclivities, which is not only to be doing mathematics as a practicing mathematician, but also popularizing mathematics: talking about mathematics to people, spreading the joy of mathematics, the insights, the games, the fun.

Many of the people for whom you write are people who are unabashedly interested, excited and curious about mathematical things. You also write science fiction, you write in a variety of different genres. But throughout there seem to be these threads of excitement about mathematics—joy of mathematics, love of mathematics—and playfulness: playing with forms, symmetries, playing with language sometimes.

I want to get back to the math-phobic individuals. Do you try to write for them in any way? Do you try to directly address any of your writings or any of your works towards people who might not otherwise have a window into this world of mathematics?

IS: On the whole not directly, in the sense that I don't sit down and think, *"OK, I'm going to write a book which will encourage this sort of person into the subject."* But indirectly, I think that if you get these people on a one-on-one situation, you can very often start to get them interested and hopefully they won't switch off when I say, *"Let* *me just tell you something interesting about how an elephant moves,"* or whatever.

I've given some talks to advanced mathematics conferences and the exact same talk to a science fiction convention, and everybody loves it. Because there are nice pictures, movies of animals, you can pull the mathematical patterns apart and everyone can see it. I think that kind of experience does change people's minds to some extent.

If I can get people in a receptive mood, then I think it is possible to get them to change their minds about what mathematics is. What I hope with my books is that at least sometimes one of these people will pick one up for whatever reason. Maybe it's because of the cover picture, or the title, or just something about it rings bells.

Quite a few people, particularly as they get older, start to think, *"You know, it's true I was never any good at mathematics when I was at school. But maybe now I ought to try to do something about it."* Sometimes they start to feel guilty that they dismissed it too easily. It's not unusual when you get older to revisit things from your childhood.

HB: Or maybe they stopped caring so much what other people think.

IS: That's right, yes. You don't mind so much when you're old, you do your own thing. And they're much less likely to be put off by somebody. If you put a 65-year-old in front of a bad teacher, he won't throw up his hands in horror and be put off the subject, he'll just say *"Oh, you don't know what you're talking about, do you?"* Older people have much more confidence and bloody-mindedness, stubbornness.

To be honest, most of the people who read my books are actually the sort of people who like reading that kind of book. It's a sort of tautology, it's obvious. The audience you get will be people who are in some way interested in the subject, otherwise they wouldn't have bought the book.

But there are some books that are exceptional. *The Letters to a Young Mathematician* book was quite deliberately aimed at a rather different audience than the usual one. That made it really fun to write. It was the publisher's idea: they had a series and they just got in touch and invited me to write the book, and I thought it was a nice idea. I wondered what I could do. It was a chance to think about the whole thing in a different way.

I enjoy writing more—and I think the book tends to be better—if I'm learning quite a bit myself while I'm doing it.

HB: So, what did you learn when you wrote this?

IS: I think what I learned was to try and put myself in a slightly unfamiliar position, which was not, *"I'm the person trained in mathematics. Here's what's going on in this subject area, let me tell you about it."* But rather more, *"You're somebody who wants to know about that sort of thing, but at the moment you don't. Now, what should I be telling you? What can I tell you that would help?"* Not just with the actual mathematical content, but with thinking about how you get it into your head.

There are various pearls of wisdom there. Something that typically causes great trouble for our students is that many seem to have the idea that you work your way through the textbook or your lecture notes line by line, in the order they were given to you, and when you get stuck you stop. You stop because you've reached the obstacle and you have to figure it out or you can't go any further.

Now what I realized, and I think my butterfly-brain helped me, was that you can actually just say, *"OK, there's a problem here, I don't understand that. I'll put it to one side, what comes next?"* And what comes next is often something that you do with the idea that you don't understand. Now sometimes if you're doing something with an idea, you figure out what it actually is. If somebody gives you a tool and you don't know what the tool is, you stare at this thing and try to puzzle out what to do with it, but if you pick it up and start hitting a nail with it, you might figure it out.

So one of the pearls of wisdom is, *Don't worry if you get stuck. Flag this as something you don't understand and move on: enlightenment might occur anyway.*

OK, you can't do this indefinitely. If it goes on too long or you've put too many things aside, at some point you have to say, *"One way or another I've got to work my way through this."* Find someone who understands it, go back to the teacher and start asking questions, read a different book: do something different. Don't just bash your head against the same brick in the wall over and over again: you're not going to get anywhere. It didn't work yesterday, why would it work today?

Look at it from a different way, you might make progress. I've run my entire life by this method.

HB: There's a lot to be said for just diving in and getting one's hands dirty. We all like to feel that we're on safe ground, that we understand things completely, but there's nothing quite like actually trying to work with a problem.

A common problem is this mentality that many people have (I suffered from this quite a bit myself, as it happens), "*Oh, I just want to make sure I understand things properly. I can't do this until I read through this subject thoroughly and this course. Yes, but I don't know everything there is to know about differential geometry or Galois theory or whatever, so I'm going to wait and go through that first.*" And often the job of a good mentor is to say "*Look: take this particular problem and try to solve it. You'll learn a lot of these things along the way.*"

I'm sure you've experienced this many times.

IS: Absolutely. I think I've been fortunate that most of my students are of the kind that very quickly works this out. But there have certainly been one or two who probably spent the first year and a half of a three-year PhD reading books, papers and courses, and didn't actually start thinking about the problem at all. It didn't matter how many times I told them not to do this, they'd still do it. They finally managed to actually make progress when they stopped doing that. And, to their credit, eventually all of them did: I've never actually had a PhD student fail—a couple came close—but they all got their PhDs and they all did it themselves.

But some took longer to work it out than others, and some knew this from day one. There was a young lady from Brazil who arrived to do graduate work with me, and during her first week here, while she was trying to find somewhere to live, she went over another student's PhD thesis and discovered there was actually a mistake in part of it and worked out what he should have done instead. She brought a stack of papers literally three-quarters of an inch thick to me with calculations doing all of this.

HB: A good choice of student.

IS: Yes, well, she was exhausting because it carried on like this. But, she was obviously used to behaving like this, and furthermore she didn't say, "*Oh, look, here's my thesis in the first week.*" She said, "*OK, now where do we go from here? I now understand where we're starting from—yes, there was a mistake but I fixed it. One case was missing, one extra possibility had to be thought out.*" And then she built on that. I've had students of that kind quite a lot.

Christopher Zeeman, who founded the Mathematics Institute at Warwick, made a point of noting that the attitude in the early days in the Math Department at Warwick was simply this: find an interesting problem, plunge in and see what happens. If after a while you're not getting anywhere, find a different problem. Maybe come back eventually to the first one—don't give up completely—but get in there and play. You will learn things. I can't tell you what you're going to learn, but you will learn things by doing that.

You might also get this kind of possibly exaggerated confidence that you can handle things. Which is helpful. Even if it turns out that you're wrong, you start out in a positive frame of mind, you start out by thinking, *I can puzzle this out. It may not be my area of mathematics, I'm going to have to learn something here, I'm not sure what it is. I'm not in my comfort zone.* But if you start to feel comfortable when you're not in your comfort zone, so to speak, I think that's very useful.

Questions for Discussion:

1. How much of what we now call "mathematical aptitude" simply corresponds to an innate sense of curiosity that has been successfully encouraged and reinforced?

*2. What role does our concern for what other people think play in our ability to learn and grow? Those with a particular interest in this topic are referred to Chapters 1–3 of **Mindsets: Growing Your Brain** with Stanford University psychologist Carol Dweck.*

IV. Mathematics and Gender

Portuguese mysteries

HB: I'd like to address the issue of gender now. There are some who claim that women can't do mathematics: "*They're not built that way. They don't have what it takes. They just don't have the mathematical disposition.*"

There are two points to be made, one of which I find to be completely reasonable and expected, and another which is somewhat confusing to me. The reasonable point is that you have empirically shown this to be completely unfounded. You have supervised many female graduate students and they have all done very well. As you say, they've all graduated and some of them have done extremely well mathematically and so forth. So there's that and that's not a surprise.

But the surprising thing to me is that they all seem to be Portuguese, somehow. How did that come about?

IS: They're either from Portugal or Brazil.

HB: Right. So Portuguese-speaking, at least.

IS: Well, it all started with a Portuguese PhD student, Isabel Laboriau, who's now in Porto. She came here and she wanted to do mathematical biology and we didn't really have anyone in mathematical biology. But I was the odd job man in the Department. She was working on the Hodgkin-Huxley nerve impulse equation and it was another one of these strange coincidences.

She was very self-propelled: she was happy to take part in actually thinking of a problem let alone working on it, she didn't need me to feed her problems. We were both trying to think of a really good problem for her to do. We'd been looking at a paper on the nerve impulse, which showed a very interesting bifurcation diagram: the amplitude of the nerve impulse varied with some parameter. And it was shaped like this: it sort of went up, did a little wiggle, then went right back down again.

Marty Golubitsky, my collaborator, was visiting at the time. And he gave a talk on his latest research work. And in the middle of this talk up came up the same diagram: identical. He was classifying a particular kind of dynamic behaviour. You didn't have to be terribly intelligent to look at this picture and look at the other picture and say, "*His techniques must tell us something about the nerve impulse equations.*" He had a general technique for finding this kind of behaviour and we were seeing this behaviour in computer simulations of nerve impulse equations: therefore his technique ought to apply. We had a quick word with Marty and the three of us very rapidly decided, *Yes, this is going to work. There's a lot of hard work to do the calculations and flesh it out, but the basic idea is sound.* Isabel got her PhD. It did work.

She then went off to Portugal, she got a job there, and started sending her good students to Warwick to do PhDs. Nearly all of them were women—there were a couple of men—and most ended up doing a PhD with me because they were interested in doing what Isabel was interested in, and that was what I was interested in. One or two worked with others. All did very well, and they came in with the natural attitude that it was perfectly normal for a woman to do mathematics. In fact, to some extent, there are probably more women doing mathematics in Portugal at the university level, and very possibly in teaching, than men.

HB: Really? There's something in Portuguese society somehow? That's fascinating.

IS: I think it goes back…in the Mathematics Department in Porto—which, until recently, was in an old building in the centre of the city—and it was also the science building and maths was part of this. There was a huge mural on the wall showing one of the great chemistry professors in his laboratory with his 3 female lab assistants. I think somewhere way back, in Portugal, this chap had discovered that women were really good at science: they were probably better than the men about actually being careful about doing what they were told, more meticulous in some ways, because perhaps women were trained to be like that.

I don't know this, but it was very striking that there was clearly a very long historical tradition of women in science in Portugal. And so, you just don't ask that question anymore, at least not in the way we would.

HB: So why is it that more people haven't noticed that? You hear every so often about these conferences and discussions on women in science. In your experience—anecdotal, but considerable given the number of instances—Portugal seems exceptional. Take a look at this place, they're doing something right.

IS: Yes. What is it? How did this happen? There is a *Women in Mathematics* organization and I think they understand this. But one of the problems for women in mathematics is that there are still not as many as there should be. I think it is improving, there are figures that show it is improving at certain levels.

But if you are a very good woman mathematician, the demands on your time are absolutely enormous. For a start, you have to get involved in the *Women in Mathematics* organization! Whereas men don't have that problem.

It's fairly clear that this is an interesting problem that needs to be investigated, this is something that is not as well known as it should be, it also goes counter to some people's prejudices, so they tend to dismiss it. But it is absolutely clear. I don't know why. But it's a fact.

HB: And we should look at it a little bit more closely.

IS: Yes. We should talk to them, find out what it is about their background. Part of it is slightly negative: mathematics in Portugal is not a fantastically well-paid job.

HB: Yes, but mathematics in many places is not a fantastically well-paid job.

IS: Well, that's true. This is something that feminists are quite rightly upset about: that there has been a cultural tendency for women to do the lower-paid jobs.

HB: In other words, this is tied to a social hierarchy.

IS: It could be tied to some negative aspects of the Portuguese structure as well as positive ones. I do remember that years ago when I was in Houston, a woman astronaut came to give a talk. She said she had been going around to high schools and explaining to all the girls that, "*You could be an astronaut!*" And she said what was happening was that the boys were being put off on the idea of being astronauts because they thought it was just a job for girls! So, you can't always win.

Mathematicians seem to have a fairly black and white view of things: if there's a counter-example, it's not a theorem. If there's a place where perfectly good, high-level mathematics is being done where women are just as good at mathematics as men, then that's that: there is clearly no innate difference in their abilities.

And there's no question that in Portugal it's like that. That means that any argument that says it's an inherent, biological effect is rubbish. Portuguese people aren't different from the rest of us. But their culture is different. So it's really got to be a cultural effect.

Questions for Discussion:

1. To what extent is it conceivable that the language you speak might somehow make you more or less attracted to mathematics?

2. What specific measures should be taken to encourage more women to produce careers in mathematics and the mathematical sciences?

V. Mathematics Everywhere

And widely underappreciated

HB: I'd like to conclude by going back to my first question about this intelligent, knowledgeable, potentially erudite individual who's afraid of mathematics. We live in a society where people talk about the dangers of innumeracy and that we should be concerned about training larger numbers of people who are competent in mathematics.

I want to take a somewhat different tack now, play devil's advocate and say, "*Well, who cares? Who cares that only a minority of people is interested in or are any good at mathematics? As long as we have this minority of mathematically sophisticated people who are being productive for society that's all that matters.*"

Moreover, some might go even further than that, saying, "*In fact, we don't have enough places for mathematicians anyway: we don't have enough places at universities. Does it really matter? Does it matter if only 5% of our society gives a fig about mathematics or thinks it's worthwhile? Society's always been like that, maybe it will always be like that, why should we even bother?*"

How would you respond to those sorts of comments?

IS: Well, I'm unlikely to respond by saying, "*Yes, that's right!*" although it would make mathematicians' lives easier in some respects. It would be a bit like the medical profession used to be in the United Kingdom with far too few doctors, which meant they got paid an awful lot for what they did. It could be to the advantage of the mathematical profession to go along with this, in some ways.

But I think where the whole thing falls apart is the perception that you only need a small percentage of mathematically capable people to run the society we're in. It's just not true. There is an enormous range of jobs where you need to know a certain amount of basic mathematics. My wife was training as an ophthalmic nurse and mathematics was by no means her favourite subject at school.

But when she was doing her ophthalmic courses, she had to know some probability related to genetic defects in eyes. She had to know basic optics, all these optical formulae (curvature of the lenses, focal length, etc.) that you learn in physics courses. She actually had to use stuff from three different areas of mathematics to do her basic training to get the qualification to move into ophthalmic nursing.

And that's a typical instance. If you start going down the list of professions and crossing out the ones you can't do because you don't know basic math, let alone anything advanced, after a while you're running out of professions.

There was a poster campaign in the United States aimed at disadvantaged people. It was telling them that understanding things like calculus empowered them. That's the same point of view and it's not just propaganda: mathematics opens up a whole pile of things that would otherwise be denied to you.

I often try to turn that question around. If you're a parent and your child is having trouble at school with math, you go to the teacher and say, "*I don't want her to do math anymore, she's no good at it.*" And he says, "*Fine, OK. By the way, she'll never be a surgeon, she'll never go into a bank, she'll never be a pilot, she'll never be an ophthalmic nurse,*" and so on.

At some point the parent is going to get really upset and say, "*But you're denying my child! I don't know what she wants to be yet. You're denying her the opportunity. You're cutting off her future.*"

Mathematics is absolutely fundamental to our culture.

There was a report for the US Government recently—I can't remember the precise details, but I do remember the figure. The net contribution of broad mathematical sciences (math, statistics, computer science, and so on) to the US economy during the first ten years of this century, in terms of enabling jobs and manufacturing industries and things like that, was 37 trillion dollars. In just ten years.

Now, this was a government report. You can look this up and find out the details, but it's massive. 37 trillion dollars in just ten years! There was one in the UK that reckoned that well over 10% of the GNP comes from the mathematical sciences: without them that 10% simply wouldn't exist.

This is actually very important to appreciate. Math is in everything we use, everything we do. If you're aware of this, from the moment you get up in the morning you start to see it. I look at the clock: it's one of these modern clocks that sets its own time by radio. Without radio, that clock would not work in that way. Well, radio works because of a whole pile of mathematics. The engineers have to know the math. We don't need to know it to read the clock, but in order to manufacture the clock somebody there has to know a certain amount about that. And that's just one example.

HB: Right. And if you improve upon the clock, if you want a better clock, a different clock, if you want the technology to keep increasing and improving...

IS: Then more and more math comes in. Our culture hides the math away. Deliberately, I think, because, if you're in marketing, you don't want a label on it saying, "*Unless you understand math, you can't use this gadget.*" This is true for any of our current electronic gadgets: iPads, Kindles, mobile phones, or GPS.

HB: I'm glad you mentioned GPS because general relativity comes into play there as well.

IS: Special relativity and general relativity and Newtonian mechanics and pseudorandom numbers and trigonometry—all of that, without which it simply would not work.

HB: It's the poster boy technology of theoretical physics.

IS: That's right. I finally got a car that's got a good GPS system in it, and I stand in awe of this thing. I did a three and a half hour journey, and it told me within about 3 minutes how long it was going to take when I started, and it was right. It was just amazing.

We have these magic gadgets, and inside them all are layer upon layer of mathematics. It may be in the form of a computer chip but the further back you go, the more math is coming in. Because as well as designing the computer chip to do the calculations, there's a whole pile of mathematics about computer chips and how to manufacture them. My eldest son works for a company that uses mathematical models to analyze how crowds will flow in big, public buildings. It's absolutely everywhere. I often feel that we should have a little red label that says "*Math Inside*". And if we did, just about everything in the world would have these labels on them.

I think people underestimate how much math is actually required, and how much our society consumes. In any given part of it, you personally may not be using this most of the time, and it's not the math you did at school. It's not that you learn math to balance your bank statement. No, the bank will do that for you. You can look at it and say, "*Hold on, I didn't pay for that*," and you can use a calculator if you're not sure how to do it.

HB: It's the **real** math we're talking about here.

IS: Right—it's the *real* math. School math is about one-hundredth of a percent of the mathematical enterprise. It's not the tip of the iceberg: it's an inch or so on the top of the iceberg. People simply don't understand just how much math is going on. And nearly all of it—maybe not straight away, but eventually nearly all of it—starts to become useful somewhere. Most of our fancy gadgets won't work without it.

HB: Well, thank you very much for explaining that to us and talking to me. It's been wonderful speaking with you.

IS: Thank you. It was a very interesting discussion.

Questions for Discussion:

1. To what extent is mathematical understanding "taken for granted" in contemporary society?

2. Do you agree or disagree with Ian's claim that, "Mathematics is absolutely fundamental for our culture"? Would you have regarded mathematics as a "cultural" activity before reading this conversation? Do you now?

Continuing the Conversation

Ian is a particularly prolific writer who has authored an exceptionally large number of highly successful books on mathematics for a general audience, such as, I*n Pursuit of the Unknown: 17 Equations That Changed the World, Professor Stewart's Cabinet of Mathematical Curiosities, Calculating the Cosmos* and *Significant Figures: Lives and Works of Trailblazing Mathematicians.* Many themes of this particular conversation were strongly influenced by Ian's book, *Letters to a Young Mathematician.*

The Science of Siren Songs

Stradivari Unveiled

A conversation with Joseph Curtin

Introduction

Finding What You're Looking For

The so-called "mystery of Stradivari" has long confused me. How is it possible, I have often wondered, that with all of the tools of modern technology and all of our detailed understanding of the way the world works, we can't seem to find a way to make instruments that can at least equal those of some fellow working in Cremona, Italy, over 300 years ago?

Not surprisingly, I'm hardly the first person to find this perplexing: it is a long-standing, indeed quasi-iconic, mystery throughout the classical music world.

It confused Joseph Curtin too. But Curtin is hardly your run-of the mill ruminator. He is one of the most recognized violin makers working today, a MacArthur Fellowship-winning artisan who combines premiere craftsmanship with an unabashed enthusiasm for acoustics and rigorous scientific methodology.

He is also a delightfully frank, open and approachable fellow.

"This idea is deeply embedded in our culture: the secret of Stradivari. As soon as you start looking at the concept from a scientific point of view, though, it starts subdividing into many different questions.

"Was Stradivari the greatest violin maker who ever lived? I would say that he probably was, so far as we know. The unknown genius is the person who made the first violin—that was probably Andrea Amati. Suddenly, from nowhere, comes this fully developed instrument. Stradivari didn't design it, so let's just say it was Amati. I'm not sure where scholarship will end up—not a lot is known. But out of Amati's

shop came these fully-formed instruments that didn't really change
much thereafter. This was in the 1500s.

"By the time of Stradivari and Giuseppe Guarneri, his great contem-
porary, in the late 1600s and 1700s, violin making had already been
going on for a couple hundred years. What they did was refine it and
take it in a certain direction, a direction that went along with the
way music was developing. That provided the players with what they
wanted: more power. They did things to the design: they lowered the
arching and changed a number of physical parameters that made
the instrument more suitable for the emerging repertoire."

The professional violinists' continual quest for more power will turn
out to be a vital part of our story, and one which is often overlooked
by the rest of us breathlessly invoking "the secret of Stradivari". In
particular, Joseph assured me, it is vital to appreciate that the bright-
ness of sound typically associated with old violins is a direct result
of continuous maintenance and improvement.

"I think what a lot of people don't realize is that if you were to take a
violin straight from Stradivari's workshop with a bow from the time,
most of the current repertoire would be unplayable. And it would
probably be inaudible too, at least in many modern venues.

"A Stradivari or any old instrument is really a re-engineered object.
It's not just the restoration—fixing all the cracks and everything else.
It's a different instrument: the set-up is so different. The bridge and
the bass bar have been redesigned. Often the top and back plates
have been re-graduated and made thinner, which has profound effects
upon the sound. And then there are the effects of often extensive
restoration. But that's completely ignored by this heroic idea of this
instrument coming mysteriously out of nowhere in the early 1500s
and then brought to perfection by Stradivari and Guarneri."

Well, that's certainly interesting. But however lovingly re-engineered
it may be, surely that's just a way of maintaining, protecting, their
uniquely compelling sound? But what **is** that sound, exactly? How
can we characterize it?

"This is very, very interesting. You might think of "Stradivari" as almost a brand, it's instantly recognizable and highly esteemed: Stradivari—old violins, great sound. Now think of other iconic music: think of Tom Waits and you can probably hear his voice. Or think of your father's voice and you'll immediately recognize something. Same thing with Frank Sinatra. Now think of the sound of a Stradivari. What comes to your mind?

*"You'll probably just hear a violinist. You may think of Itzhak Perlman—he plays a great Strad. Maybe you hear Heifetz, or whoever your favourite player is. You hear **their** sound, and they play a Strad or a Guarneri. But let's try and track it back a little bit: how much is the violinist and how much is the instrument? The question becomes, Can the **players** tell the difference? That was the radical thing with the blind-listening test."*

And so Joseph, together with his inquisitive colleagues Claudia Fritz, Jacques Potevineau, Palmer Morrel-Samuels and Fan-Chia Tao, decided to conduct an experiment in 2010. They selected 6 premiere violins—3 classical greats (two Stradivari and one Guarneri) and 3 contemporary instruments—and led 21 expert violinists who were gathered for the 8th International Violin Competition in Indianapolis to a hotel room, put goggles on them so they couldn't see which instrument they were playing, and asked if they could tell the difference.

The results were astounding: the majority of these expert players unknowingly preferred a newer model.

"There was a clear favourite, which happened to be a new violin, and there was a clear least-favourite, which happened to be a Strad. In the middle there were violins that certain people loved, others hated, and the rest held in qualified regard."

This naturally raised a lot of eyebrows. But then came criticisms of the experiment. What can you determine in a hotel room? And why just ask the violinists themselves? Why not judge things properly, using expert listeners sitting in a proper hall?

So Joseph and his colleagues did a follow-up experiment again in Paris in 2012—a much more comprehensive set-up that involved 10 soloists with 12 violins (6 new and 6 old), professional listening environments, expert listeners and even an orchestra.

What did they find? Well, pretty much the same sort of thing. Neither the soloists nor the listeners could tell the difference between new violins and old, while something like 6 out of the 10 violinists preferred the new violins.

So what does all of this tell us about "the secret of Stradivari"? Perhaps simply that it might be more of a reflection of what we'd *like* to believe than anything we actually should.

  "I remember a conversation I once had with Gabriel Weinreich, one of the top physicists who went into musical acoustics research. I had a theory of why this or that happened, and he said to me, 'You know, before you start making theories to explain a phenomenon, it's best to make sure that it actually exists.'

  *"You can look around forever for the secret of Stradivari, but unless you can actually **define** what you're looking for, you're not going to find anything."*

The Conversation

I. Finding One's Niche

Becoming a violin maker

HB: How does one actually become a violin maker? How does that happen? My understanding is that, in your case, it happened in a somewhat different way than it happens to most. Or perhaps there is no general way...

JC: Generally speaking I'd say it's a circuitous route. It's not a profession that your high school or college career counsellor is going to suggest to you.

As a boy, I would never have dreamt that I was going to be playing the violin—music had never crossed my mind. Nor would I have dreamt that, trying to be a violinist, I would eventually become a violin maker. You get exposed to these worlds and then you start to see that they're interesting.

I fell in love with violin playing, but I started rather late and I didn't have good teaching, so it ended up being very frustrating. I started playing at an English grammar school I attended; and then when we moved back to Canada I played in the high-school orchestra. I didn't get any private instruction. It was only when I went to university that I started taking private lessons. Of course I improved, but then I was told to quit by a famous violinist, Lorand Fenyves, who was one of the best violinists in Canada—one of the most sought-after teachers, anyway.

HB: He told you to quit? That's pretty harsh.

JC: Yes. He invited me to his master class in Banff and then told me to quit, which was particularly painful. I could see his point, though.

I mean, he sees this 19-year-old who obviously loves it, but is technically ill-equipped. It was really very difficult to be told that you can't do the one thing that you want to do.

But I kept taking violin lessons. One summer I was working as a busboy and I decided that I was going to buy a better violin. I got the name of Otto Eredesz, who was a new violin maker in Toronto at the time, and his wife Rivka, who is an Israeli viola soloist.

I visited them and ended up staying for several years. She taught me viola. It was then that I realized that—well, if you want to be a French poet it's not enough to be poetic: you've got to speak French. That was the trouble with me and music: it wasn't internalized. I know many top musicians now and I've seen how their kids are brought up: music is a mother tongue, in a sense. I can't really carry a tune.

HB: Do you think this is something innate? Or do you think your situation would have been different had you had better or different teaching from an earlier age?

JC: I think that, if I had started at an earlier age and loved doing it, I would've progressed to whatever level I could have reached. But what they're learning more and more about the brain's plasticity is that you can do that at any age. Indeed, a couple years ago I got interested in practicing again, and I started taking more lessons.

I should say that I only took two lessons, but it was great. I really progressed. Within a month, I was better than I had been 20 years ago. I solved so many problems that I had previously thought insuperable, just by being able to analyze the problems and not getting hung up on being a violinist. It was a very satisfying period, and now I play much better than I did before.

But when I was younger there was something about my ambition: I wanted to be a violinist, but I wasn't that interested in the fine details of playing.

As soon as I got into making violins, however, the details attracted me immediately. I realized that people who are good violin makers are just enormously interested in how bits of wood look, how one curve works versus another and how different things sound.

And the people who do well as players or singers, I think they are the people who are really interested in the details of their craft, interested enough to practice them. But, in fact, I didn't really do that. I loved the emotional effect of playing. But I understand all that better now and I think I probably could have been a decent violinist.

What really stopped me playing was that it started messing with my ear. Players get ear damage over the years. It's really a problem for musicians.

HB: Is this tinnitus or something?

JC: It's a hearing loss related to high levels of sound. A violin from 3 feet away is typically 90 decibels, but from right below your ear it can be over 100 decibels. If you do that for hours at a time it will wreck your hearing. My ear would start to ring a little.

HB: Perhaps it's because you make violins that are too good. Maybe if you made some poorer quality violins that didn't project as well...

JC: Well, more powerful violins obviously make a difference. But this actually underscores the fact that violinists really need to take measures to protect their ears. Some use earplugs, which I don't find very satisfying. But you can also put a little mute on your violin; it doesn't have to be a full mute. Players can take sensible precautions like that. One's hearing is a perishable commodity, unfortunately.

I met a violin maker in England who had two hearing aids, which really struck me because it was just at the point when I started worrying about my own hearing.

He said to me, "*You know, if you think you're going to need hearing aids, it's better to get them early, because most people get them too late.*"

I think of my parents who have both lived to good ages; my mother is 92 now but she needs a hearing aid all the time, and the same with my father. But people start too late with them and your brain can't figure it out, so it's just a nuisance and you don't end up putting them in. However, if you start when you're still in good shape, then you can adapt to it very well and it becomes very useful.

As a violin maker, one needs to hear to discriminate. I'm still doing fine but I don't want to squander it away.

HB: Let me get back to the beginning of your life as a violin maker. My understanding of your story is that you were struggling for a bit, you played the viola with this Israeli woman whose husband was a violin maker, and then you somehow learned how to make violins from him.

Now, I'm not a particularly handy person, but it seems to me that you don't just fall out of bed one day and decide that you're going to make a violin—or perhaps you do? How does that actually work?

JC: I can only speak for myself. My earliest passion was trying to make things. I was very interested in trying to figure out how things worked and designing them, although I didn't have any knowledge of tools. I didn't have a father who had a tool shop and could show me. I don't think I finished anything until I finished my first violin. I tried making tape recorders. I would buy the spools of tape and I would read up on how they worked, but then I realized how hard it was, and that I couldn't make it. I wouldn't even acknowledge that; I would just move on to something else, say watches, and then I would try to design watches.

When I was 14, I got very interested in violin making. I had a book that a lot of violin makers start with by Ed Heron-Allen, a charmingly-written Victorian book (*Violin-Making as it Was, and Is*).

And I thought, *I can do this*. There were instructions, and there was even a drawing of a violin you could cut out. So I went down to Heinl's in Toronto, a violin-making shop, and I bought some parts. I think I spent a few days trying to inlay bits of clam shell in the pegs: that was my idea of what was important. I botched one and put it aside. I got the spruce—you have to glue it together—and I got my dad's plane, which was the tool for the job, but it was as blunt as a butter knife. I put that aside and I got a rasp, but everything was just so shockingly glued together that I lost interest and moved on to something else.

I thoroughly put that behind me, but when I met Otto I saw that here was someone who was actually making a living at it. And I think that he saw that I was interested: he saw it in the questions I asked and the things I noticed. How can you tell if someone can do something? If they're interested. I would hang out with him and ask questions and notice things. After a particularly discouraging summer, when I had been playing chamber music and feeling that I would never play violin professionally, I decided to ask Otto if he would teach me how to make violins. And the very day I was going to ask him, he suggested it to me. It was almost spooky, his timing. I guess he sensed that it would be a good path.

Back then I thought, *Violin making? Who would want to be stuck up in a room with a bunch of sawdust for the rest of their life?* It didn't seem romantic at all. But whereas violin playing seemed like a laby-rinth—I didn't quite know how to get through and I couldn't really tell how good I was—violin making I could see. I could figure it out. It was just like an open door. It would be really hard to learn, but I didn't see any obstacle in the way, any real impenetrable barrier. So I took to it, and Otto was very encouraging. That made all the differ-ence. If I hadn't met him, I can't imagine quite what I would have done.

A friend of mine told me that when he was trying to figure out what he was going to do, he went to a career-counselling service in Chicago, where they give you batteries of tests for three days.

They thought that he should be an engineer and he had never thought about that before. He had thought about being a private detective and various other things. Now he's been an engineer for many years and still does a lot of other things. But what I remem-ber him telling me was that they told him that if you have a gift in one particular area and a job in another area, you're not likely to be satisfied. You might have it as a hobby, but you want a career that overlaps with the things that you're good at and the things that you're interested in or passionate about. For me, the amazing thing about violin making is that, if I look back at my life, the things I've been truly passionate about are making things, science (I always imagined I'd be an electrical engineer or something like that), music, and the visual arts.

I think of myself as someone who flits between subjects. I don't tend to finish things—I tend to keep coming back to them. With violin making, if I get bored at the workbench I can go up and do some acoustical measurements.

HB: But you *do* finish the violins, I'm guessing. I mean, if I were to commission a violin from you today, it would get finished at some point, right?

JC: Yes, that's been the saving thing about it. I pretty much work on commission: I've got to get it done and that's that. It's an interesting concept—I think it's from the Myers-Briggs personality test—whether you're a "closer" or a "finisher". I've realized that I don't tend to finish things, whereas my wife is an incredible finisher. I think couples may be drawn together that way. I think it balances things out. People who finish things quickly maybe should have done some revising, and people who never finish things—well, they should work on that too.

I've had violins come back to me years later for an adjustment and I'll re-varnish them. Nothing ever feels done. I'm the same with writing: I tend to keep revising. But as long as there are deadlines where things have to officially go out in the world, everything works out. I don't think of it as a bad quality, just one that needs to be under control.

HB: So, you developed these skills and then you went out into the world to make your living as a violin maker. I could imagine, not knowing anything about this, that it's awfully difficult to get started, to bootstrap things. You're young, you have some skills, but you certainly don't have a name because you're just starting out. How does that work? How do you start getting a name and getting commissions to move forward when you're brand new in the field?

JC: I'll tell you what I think is a standard route nowadays: you study at a violin-making school, and then you go to a restoration shop. You apprentice there and learn to work with old instruments, and then maybe you set up on your own.

Otto was an informal person. I remember that, after I finished my first instrument at his place, for some reason I was really quite rude: I suddenly decided I had to make one entirely on my own. So I disappeared, without calling. It didn't occur to me that that was not an appropriate thing to do. I made my second one entirely on my own, and then I brought it back. After that I would go back and forth a lot to Otto's.

I sold a few violins around Toronto but then I got some contacts in New York somehow. My girlfriend at the time was living in New York and she was a violinist, which helped. So I started selling instruments in New York quite early on and that got me through. If you sold a violin back then for a few thousand dollars, that could last a few months. It wasn't bad. My father co-signed a loan for me back then for $5,000 and I like to say that I only paid it off last year.

HB: I'm trying to get a sense of the market. I'm thinking to myself, Who is going to buy a custom-made violin? If I'm just a normal music student, or even a fairly accomplished violinist, would I be thinking about buying a custom-made violin? The market for custom violins is probably not very large compared to the market for those you'd find in a store. Is that right?

JC: When I was starting, my teacher Otto and a few other makers, particularly an American named Sergio Peresson, were the first people to really be making their livings as makers rather than maker-repairers or maker-dealers.

I didn't really get started until I met Gregg Alf in Italy. We came back here to Ann Arbor in 1985 and started our company, Curtin & Alf. Our two minds complemented each other's rather well: he had some good market instincts and I had some sort of anti-market instincts. I remember my first time trying to sell a violin. I went to my old university and showed it around, and a friend of my sister, who was a player there, said, "*I don't understand. Is that instrument actually **for sale?***" I had a compulsion to point out everything that was wrong with the instrument, so that they wouldn't think I didn't notice it. There's a whole psychology of presenting yourself in the world, and it takes a while to build up confidence. But Gregg and I together could do more than either of us could do individually.

So we started a firm. Ruggiero Ricci was a world-famous violinist who was living in Ann Arbor at the time, and some of his students bought instruments. Then Ruggiero commissioned a copy of his Guarneri.

That was really a breakthrough. He let us use his photo as an advertisement and we put full-page ads in *The Strad*, which hadn't really been done by makers before. Then the phone started ringing. But it was pretty touch-and-go for a while.

Questions for Discussion:

1. *Are you surprised to hear that so many professional musicians and those working in the music industry have hearing problems? Do you think that should be more widely acknowledged? Might there be an aspect of pride preventing people getting hearing aids earlier? (Those interested in more details of how neuroplasticity plays an integral role in our ability to interpret sounds through artificial hearing devices are referred to Chapter 9 of **Knowing One's Place: Space and the Brain** with Duke neuroscientist Jennifer Groh.)*

2. *Does anyone have the potential to become a world-class musician if exposed to music from a sufficiently young age?*

3. *Are apprenticeships sufficiently appreciated in our society?*

II. Confronting Biases

Measuring a mystery

HB: I'd like to talk in a bit more detail about these famous 17th—18th century violins now. You've made some copies, as you've said, and you've done considerable acoustical research, while rigorously investigating a number of innovative directions: new types of violins, new sounds and new materials.

But before we get to that, there is something that I'd like to address: this notion of the "ultimate violin", the Stradivarius. The commonly-held belief is that these guys in the 1720–30s were able to produce violins of unparalleled quality and resonance, with a sound that no one has been able to duplicate since. There is this notion that this time was somehow the apex of violin-making; and in some bizarre way, notwithstanding all the modern technological advances we have today, we haven't been able to replicate it. Coming from a scientific background, I find it odd that we wouldn't be able to do that.

JC: Yes, this idea is very embedded in our culture: *the secret of Stradivari*. As soon as you start looking at this concept from a scientific point of view, though, it starts subdividing into many different questions.

Was Stradivari the greatest violin maker who ever lived? I would say that he probably was, so far as we know. The unknown genius is the person who made the first violin—it was probably Andrea Amati. Suddenly, from nowhere, comes this fully developed instrument. Stradivari didn't design it, so let's just say it was Amati. I'm not sure where scholarship will end up; not a lot is known. But out of Amati's shop came these fully-formed instruments that didn't really change much. This was in the 1500s.

By the time of Stradivari and Guarneri, his great contemporary, in the late 1600s and into the 1700s, violin making had already been going on for a couple hundred years. What they did was refine it and take it in a certain direction, a direction that went along with the way music was developing. That provided the players with what they wanted: more power. They did things to the design: they lowered the arching and changed a number of physical parameters that made the instrument more suitable for the emerging repertoire.

Now, I'm not a person who is very knowledgeable about violin-making history. But I think what a lot of people don't realize is that if you were to take a violin straight from Stradivari's workshop with a bow from the time, most of the current repertoire would be unplayable. And it would probably be inaudible too, at least in many modern venues.

HB: Really?

JC: Yes. A Stradivari or any old instrument is really a re-engineered object. It's not just the restoration—fixing all the cracks and everything else. It's a different instrument. The set-up is so different. The bridge and the bass bar have been redesigned. Often the top and back plates have been re-graduated and made thinner, which has profound effects upon the sound. And then there are the effects of often extensive restoration.

But that's completely ignored by this heroic idea of this instrument coming mysteriously out of nowhere in the early 1500s and then brought to perfection by Stradivari and Guarneri; and then...

HB: ...declining ever since?

JC: Yes. What I'd like to research is who actually started instituting these changes. They started using longer fingerboards, because people wanted to go higher up. They put the neck in differently. Various people are figuring out who did what when, but as far as I'm concerned these are equally important contributions. They just don't have a glamorous story associated with them.

HB: But there's still the question of the sound, right? Again, these famous, world-class violinists, an awful lot of them, *do* seem to be playing a Stradivarius. Isn't that so?

JC: Many fantastic players will say that the old Italian sound is the best, and it's what we're all looking for.

HB: OK, but how do you characterize that scientifically?

JC: Well, this is very, very interesting. You might think of "Stradivari" as almost a brand, it's instantly recognizable and highly esteemed: *Stradivari—old violins, great sound.* Now think of other iconic music: think of Tom Waits and you can probably hear his voice. Or think of your father's voice and you'll immediately recognize something. Same thing with Frank Sinatra. Now think of the sound of a Stradivari. What comes to your mind?

HB: Well... maybe it's because I'm not an expert, but I just think of a violin.

JC: Exactly! You'll probably hear a violinist. You may think of Itzhak Perlman—he plays a great Strad. Maybe you hear Heifetz, or whoever your favourite player is. You hear their sound, and they play a Strad or a Guarneri. But let's try and track it back a little bit: how much is the violinist and how much is the instrument? The question becomes, *Can the players tell the difference?* That was the radical thing with the blind-listening test.

I initially assumed, like virtually everyone else, that there was an "old sound" that came with age and that, even if it wasn't best exemplified by any particular maker, there was at least a general quality there. And God knows the amount of time we spent looking at old wood samples and trying to find differences, but researchers *couldn't* find physical correlates with these different instruments: if you look at response curves, sound radiation or emittance, you see similar sorts of things. All violins are different, but whether new or old, regardless of what they look like, they all have similar features.

And some people would say, "*Oh, we're not looking at the right thing: we've got to find a different measurement.*"

But there is always the possibility that maybe there really *isn't* a difference. And what do we mean by saying that there *is* or *isn't* a difference anyway? What we're talking about is something that can only be heard if it's being played by a player. So if a *player* can't hear the difference, and the *listeners* can't, then you have to say that there's objectively *no difference*. And that is **testable**.

HB: This very much resonates with my confusion. I would hear these stereotypical notions that there was this wonderful violin, the Stradivarius: it was the apex and nobody could duplicate it. This is the standard story, the cliché, as you were saying. And I'd think to myself, *They must be able to figure out what the relevant parameters are. We have all this advanced technology, we have all these different methods of testing; we must be able to determine whether it was the type of wood, the type of resin, or whatever. If the sound quality can be parameterized in a clear way, we should be able to identify those things*. The idea that it was somehow irreproducible was very mysterious to me.

JC: I remember a conversation I once had with Gabriel Weinreich, one of the top physicists who went into musical acoustics research. I had a theory of why this or that happened and he said to me, "*You know, before you start making theories to explain a phenomenon, it's best to make sure the phenomenon actually exists.*"

So you can look around forever for "the secret of Stradivari", but unless you can *define* what you're looking for, you're not going to find anything.

You might say, "*A violin's sound is subjective: how can you measure that?*" Well, you get a bunch of them and you start to find statistical trends. The obvious way is to use a double-blind test. It's the only way you can get rid of subjective biases: if you know that a particular violin is a Strad then you're going to hold it differently, you're going to treat it differently, and listeners are going to be changed by that knowledge.

HB: So what happened? Tell me about the double-blind test. What were the results?

JC: I started working with Claudia Fritz, a French researcher I met at Cambridge—we have a group of violin researchers who meet there. She was working with Jim Woodhouse, a lead researcher and Cambridge professor studying how we perceive violin sound: how we judge it and how people evaluate instruments. And she started doing some blind tests. We invited her to the VSA Oberlin Acoustics workshop, which is a week-long workshop in Oberlin, Ohio, that I direct with Fan Tao, who was also a part of this study. So we suggested to Claudia to come to Oberlin and do a blind test there.

There were 12 violins on a table and various people would put on goggles and play in a darkened room. Then they would arrange the violins in order. It's very interesting what happens when suddenly a normal source of information is taken away.

It's disorienting, but there's also a sort of freedom—there's this curious mixture of uncertainty and liberation as the primacy of your perception forces you to really listen. It impressed upon me that this was the way we should be studying the difference between new and old violins.

There's a big violin-playing competition in Indianapolis and Glen Kwok, the director, is a very open-minded and interesting man. That was an obvious place to do tests because there were a lot of great violinists there and a lot of good violins. So I started organizing that and invited Claudia to that as well. We designed the experiment based on what we had learned in Oberlin and experiments Claudia had done in Montreal with people at McGill University.

HB: How many people were involved there?

JC: We had 21 subjects and 6 violins: 3 new violins, 2 Strads, and a Guarneri del Gesù. We did it in a hotel room. Claudia was 6 months pregnant, and we just went around rounding people up after concerts—we were just trying to pull something off. But we did it well, I think; and it was so striking watching people. It's one thing reading about it, but when you sit there and watch very good players pick up instruments and then ask, "*Which do you prefer? A, B, or C?*" And they start picking the new ones without knowing. And you don't even know it when you're there, because you can't see them either—it's a double-blind experiment. But afterwards you find out what happened.

HB: So describe to me how it works for the violin players. How is that knowledge screened from them?

JC: They wear goggles while playing. You turn the lights down, so the violin is just a silhouette, and you really can't tell one from another by sight. We used some sort of essential oil to mask scents.

The players were not completely blindfolded because that disorients you more profoundly, as Claudia and her groups had found in other studies. So they just wear goggles instead. Then they had 20 minutes to choose their favourite, rank them all, and then talk about the different qualities of each one.

A number of things were very clear to Fan and me just from listening. Fan is an amateur violinist and has a good ear. The biggest difference was the violinist. I remember a Russian violinist with a great bow arm made them all sound like a late Guarneri del Gesù; with another player they all sounded kind of bright and hyped up. It was very impressive: it was largely dependent on the violinist. And the violinists had strong feelings about the differences between the violins. It wasn't as though the conditions took away their ability to discriminate. They could discriminate very thoroughly.

The results were very interesting. There was a clear favourite, which happened to be a new violin, and there was a clear least favourite, which happened to be a Strad. In the middle there were violins that certain people loved, others hated, and the rest held in qualified regard. This one particular Strad was considered both the easiest to play and the hardest to play, by an equal number of people.

So there's a large middle group. You're not going to learn a lot about what players like by measuring the middle group. You need to take the ones at the extremes: take 10 violins almost everyone likes, and 10 that almost no one likes, and measure them. That's when you'll start finding something.

HB: That harkens back to the comment you were just talking about that your friend Gabriel Weinreich made: you have to know what the phenomenon actually is, what it is that you're looking for.

JC: Exactly. A lot of other people have measured violins and have even derived "quality parameters'", i.e. physical correlates with subjective judgements. But throughout it was kind of assumed that violin quality was obvious: you just have to give it to a good player and ask them.

Sometimes you see it written, "*Any violinist can tell within 30 seconds whether a violin is new or old.*" This is something that was just taken for granted. I thought the same. And these were physicists!

HB: Well, you know about those guys...

JC: Well there's an important point there. Physicists work in labs. You don't get Strads just showing up in labs, so a lot of this work is done using very poorly made "violin-like" objects. This doesn't matter if you're trying to work out the basic physics, but if you're trying to talk about quality, then you've got to start involving good players and good instruments.

Now, unfortunately, I think there's been a general anti-science feeling in the violin-making world and even in the music world. I can't tell you how often I used to hear that there's a kind of pride in the mystery of the violin.

HB: So it's as if you don't want to go there? Like learning the secret behind a magician's trick?

JC: Yes. Some people think, *Why would you want to lose that mystery?* Of course, that's not how it works: the more you understand, the more interesting something is.

HB: And the more you discover you *don't* know, of course.

JC: Exactly—it never stops. I would say that violin makers weren't very interested in science in the past. There's a wonderful violin researcher named Norman Pickering, who said to me once just as I was about to give an acoustics lecture, "*You know Joe, there's a special silence that comes at the end of an acoustics lecture for violin makers.*"

It's true. It's just out of the usual context, really. And they could stay within this region of "working by intuition," because it was "well-known" that science couldn't explain it.

How many times do you read media headlines that say something like, *For centuries, scientists have struggled in vain to understand the secrets of Stradivari...* Really? **In vain**? Do you have any *idea* how many interesting things have been learned? It's like saying that scientists have struggled for centuries to understand the human brain. In vain?

HB: And there's been a lot of progress.

JC: Yes. And it's still very interesting and still a mystery. That's just how the world is. Now, if you were to say, "*Scientists have struggled in vain to cure cancer...*" Okay, well in that case you know what the answer is: either you can cure it or you can't.

HB: Even still, we've made a lot of progress.

JC: Absolutely. And I know nothing about cancer, but at least there's a clearly defined goal: to ensure that people don't die of cancer. Whereas with the secret of Stradivari, *what* are we talking about here? So much energy was wasted just in looking for the wrong things.

So back to the Indianapolis experiment. It got a lot of attention and a lot of criticism, but very little from scientists actually. The criticism from the violin world was, "*What can you figure out in a hotel room? You can't determine anything in a hotel room.*"

Well, hang on a minute: do violinists play in hotel rooms? All the time! In fact, the better they are, the more likely it is that they do, because they live in hotel rooms when they're touring.

But the point is nonetheless a fair one. Projection, which is how well a violin can be heard at a distance, can't really be judged from under the ear. It can only be judged by a listener—and we didn't have listeners.

HB: Right. So this speaks to the need for more tests.

JC: Yes, exactly. It was a small test but it provided a kind of counter-example to the received wisdom. The next test was much bigger. That was organized in Paris by Claudia. She got a grant through her lab.

HB: When did this happen?

JC: September 2012. It was a week long. We had 10 soloists with 12 violins in total: 6 new and 6 old, 5 of which were Strads. We thought, to ourselves, *Okay, last time the 3 old instruments we had in Indianapolis didn't happen to do as well, but let's try again with these 6. Nothing is conclusive.*

Of course, the trouble with violins of this quality is that we had $100 million worth of these instruments, which equates to only a few. You can only really learn by doing a lot of small tests—we'll never be able to do something as big as a drug test. I mean, there are 600 Strads in existence. I suppose it's conceivable you could get all of them together...

HB: It's logically possible, but probably unlikely. OK, so what happened? You did this test and what were the results?

JC: Well, pretty much the same. We gave the soloists more time. They were in a rehearsal room. It was more like a normal room that you would practice in—it wasn't a hall. They had an hour with these 12 instruments and we asked them to eliminate the ones they didn't think were suitable, which they could do fairly quickly, and then focus on the ones they wanted. In the end, 6 out of 10 chose a new violin.

HB: Just out of curiosity, were any of these new violins yours?

JC: Nobody will ever know who made any of the new violins. Not only was it double-blind, it was also set up so that it would not be a marketing opportunity for any particular maker. They were all professionally-made by 6 different makers.

Anyway, on day one, 6 of 10 chose new violins. But the next day we did it in a hall: a beautiful, small hall with excellent acoustics, together with a piano accompaniment. The players could hear how it felt with the piano, and they could also have a trusted listener out in the hall to give them feedback, or another soloist to play alongside. So they did all the things a violinist would normally do, except for the fact that they didn't know what instrument they were playing.

HB: Did that change anything? What were the results then?

JC: It was pretty much the same. The next day we had just a few violinists playing excerpts of violin solos. They alternated, so that you could listen to two violins and compare them. What we were interested in was projection: which can you hear better, which is clearest? We were interested in the violin's projection both on its own and with an orchestra. We had a whole orchestra assembled. We chose excerpts from the climaxes of different concertos where the violinist tends to be almost drowned out by the orchestra. We also had an audience of about 50 experienced listeners composed of violin makers, players, and acousticians.

After an hour or two, there was no question about it: the new instruments projected better. Old instruments tend to sound softer under the ear but there's this notion that they project well nonetheless, that somehow the sound seems to grow the further it gets from the instrument. This is a commonly discussed thing: "the mellow instrument that projects". It's a mystery. But, again, does it exist?

There was a particular Strad that was very well-liked—I believe it was the third most preferred instrument. But it didn't project very well. It had a lot of energy at the low frequency, but to project well you need to make a lot of sound at high frequencies. Your ear is most sensitive at about 3.8 kilohertz; 2–4 kilohertz is the range where we pick up a lot of things. If there's not much energy, you're not going to hear it very well. That's what seemed to be the case here.

Under your ear, on the other hand, the mellow sound can be very enchanting and lovely to play. I was surprised that so many people liked the violin that did best in this test because it was quite loud under the ear, almost to the point of being irritating.

HB: Wouldn't that go against this whole phenomenon of this "enchanting, mellow sound"?

JC: Well, there's no doubt that loudness from under the ear varies from instrument to instrument. A caricature of new violins is that they are bright and loud under the ear and kind of crass, while the caricature of old violins is that they are very mellow and smooth. But there are actually some old ones that are very bright and very loud under the ear.

HB: I'm thinking that if the signal is that distinct, if that caricature is accurate, then the people playing would be able to tell. They should be able to say to themselves, "*This is a mellow violin, so I must be playing a Stradivarius*," or something like that.

JC: Well, this is what surprised me. At least for soloists, they seem to go for big sounding instruments. I did acoustical measurements for all the instruments: I was measuring the sound output all around the instrument—we can talk about how later on—but it's the sound radiation per unit force at the bridge. There was a definite correlation: the ones that the soloists preferred put out more sound. Right across the board. If you plot the average output levels of the two most preferred, it's about two decibels above the two least preferred.

This hadn't really been isolated before, and the reason is because we didn't have such a clear selection process: a lot of violins are just right in the middle and they tend to blur the data.

Now what's really interesting is that we also had the players, unbeknownst to the audience, play an excerpt of a concerto on their own instruments and half of them were new violins and half were old.

Then we asked the audience to listen to 30 seconds of this concerto and guess if it was new or old.

The first two that were played, the majority of people believed one way or the other, but in both cases it was wrong. The rest was completely random. Also during day two, after the soloists had been playing in the hall, we presented each player with a series of instruments and gave them 30 seconds to determine whether it was new or old. Again, the results were completely random.

People would say things like, "*Oh, this is probably from the 19th century.*" They were hearing something distinct.

So this doesn't mean that there *aren't* any differences, but we can clearly conclude that our findings showed that neither the player nor the listener could tell old from new. Again, it doesn't mean that there aren't any differences—indeed, there clearly *are* differences, because if they simply had said that all the violins they couldn't hear as well were old, then they probably would have gotten it almost right.

That's the crazy thing: there *are* differences, but that's not what people are listening for. The thought was that old instruments project better, or that they do this or that.

HB: So these preconceptions that we impose, biases that are actually quite distinct from what the reality is, have a significant influence. As you said, if they had just concentrated on the projection then they would have been able to distinguish between the new and old instruments. But they're bringing a different perspective when they play, or when they listen.

JC: Yes. When I play a violin now, even knowing everything that I know, it's really hard for me to subtract the impressions. If I pick up an old violin, I have to remind myself that what I'm hearing is what hits my eardrum as well as everything I know and love about these old instruments. When I play my own instrument, it's what hits my ear combined with my own self-doubts and worries. All these things colour my impression of the sound. And all of this makes it important to start objectively measuring sound.

I like to evoke the image of a pilot: when pilots are flying and the weather gets bad, they don't trust their instincts, they trust the instrumentation. In an ideal world, instincts and measurements work together, but right now violin makers are used to measuring things, but not the sound itself. The equipment hasn't been made available.

There was another general difference between the old violins and the new ones besides the amount of raw power being produced, which is that the older ones are more thinned out in general, I think you could say. Most new makers wouldn't go as thin as that.

HB: So what are we talking about here, exactly? The thickness of the materials used?

JC: The thickness of the top and back especially. As you thin things out, the frequency of the resonance goes down and you get a slightly darker sound. This is a generalization, but there's a trend there.

This is something that requires a lot more study, but in looking at the characteristic curves of those 12 old and new instruments, there was a general difference that I think could be explained by a difference in average thickness. You don't have to go to properties of old wood or anything elaborate like that to explain this phenomenon.

So that's what we've found so far. A similar test was done again in New York, with a smaller blind test, again studying projection. And again the results were pretty much the same: the audience couldn't tell and the old violins didn't project as well.

HB: So to sum up, a couple of interesting results seem to have come out of this. One is that the players and the listeners actually prefer the sound of the new instruments over the old ones, but they can't clearly identify which instrument is new and which is old while they're being played. We know that there are measurable differences between new and old instruments in terms of the quality of the sound—there is actually a way of measuring some of this. We don't know exactly what produces these differences but you have your suspicions. Further-more, even though there are these objective differences, the people who are listening and the people who are playing cannot accurately identify which instrument produces these characteristics. Because they have these preconceived, subjective notions that they bring with them when they start to play or listen.

Questions for Discussion:

1. Were you surprised by the results of the experiments that Joseph describes in this chapter? If so what, in particular, did you find most surprising?

2. To what extent do you believe that a phenomenon is less beautiful if it is fully understood?

III. Characterizing Sound

Towards objectivity

JC: This brings me to something that I'm greatly interested in, developing the ability to describe what we're hearing in objective terms rather than subjective ones. One of the difficulties for violin makers is that violin players will say things like, *"I'd like the sound to be a little darker"*, *"a little brighter"*, *"a little warmer"*, *"a little woofier"* or whatever. Some of them are clear enough, others much less so.

What would happen if we just described the objective characteristics of sound? Well, you can't do that unless you can hear them clearly. Musicians often have incredible ears; you can play a bunch of chords and they can tell you what notes you're playing. They're used to hearing analytically, but usually just in musical terms: the notes that make up the chord or the musical structures.

There's no reason why we can't learn to hear timbre in those same terms. Recording engineers do it all the time. They train by using software that tells you, for example, if a sound is at 2 kilohertz you'd be in the third octave band. There are random changes that they have to identify. You can gradually train your ear to hear differences and explain those differences. For example, *"Well, there's a bit of a boost at 3 kilohertz."*

HB: Do you think that some people may feel that if they train or retrain their ears to hear in this particular way, that will somehow interfere with their musicality? That they'll somehow begin thinking too clinically, too scientifically? That they'll start thinking in a way that prevents them from playing music at the highest possible level? Have you heard those sorts of arguments?

JC: Right now ear-training for violin sound doesn't exist, so that issue hasn't come up. I've been thinking about it and I've given the occasional talk about ear-training, but I can't imagine why anybody would feel that way. Musicians are incredibly pragmatic people. They spend hours simply playing scales. Does that negatively affect their musicality? No.

HB: Sure, but we talked a little bit earlier about this sense that some people have that you don't want to be too technical or too scientific because that will somehow "ruin the mystery". So I'm trying to imagine an argument where people might say something like, "*This is all very clinical and scientific and may interfere with my playing or give away the mystery.*" Anyway, I'm just speculating. Apparently you haven't heard any concerns like that.

JC: Not yet, but I'm sure it'll come up. There's always a reason not to do something. Personally, I think this will clarify things. As I said, recording engineers are doing this all the time—they don't even talk about it; they just make the adjustments. I would think that becoming better acquainted with reality and identifying it more clearly is a good thing.

HB: Let's get to some of your techniques for analyzing and hearing these sounds. You've developed some of these techniques yourself, and presumably there are other people who've developed methods as well.

JC: Yes. I've had an ongoing project of measuring violin sound in a workshop setting rather than in a laboratory. I've taken techniques developed by others and adapted them and refined them for violin-making use. I find, as a working maker, if you can't measure something in 10–15 minutes, you're not going to do it. My definition of boredom is watching a physics experiment in progress: setting up all day, testing software, connecting all these different parts together. And then—boom—it's over. That's not going to work for violin makers.

So let's think about what's happening in a lab. In a lab, there's a saying that equipment is over-designed if you can use it more than once: they generally design something for a single experiment and then move on.

But I wanted something that would evolve into a tool that could be used over and over again to measure violins in a standardized way. It would be a workshop tool, a measurement tool, rather than a physics tool—although hopefully it could be used for that too from time to time.

A German violin maker and researcher called Martin Schleske started using what's called an impulse hammer or an impact hammer. It's a tiny little hammer with a force sensor in its head. It's a widely used engineering tool, particularly in civil engineering. You can whack something like the Golden Gate bridge with it and put accelerometers all over the bridge to study its normal modes and examine its stability and so forth.

I think it was in the 70s that some very small impact hammers started coming out that could be used for tapping violins and violin bridges. And Martin Schleske, whom I got to know quite well and admire very much, had it set up so the violin was held upright. You'd tap the bridge and rotate the violin with respect to a microphone so that you could measure the sound all around it.

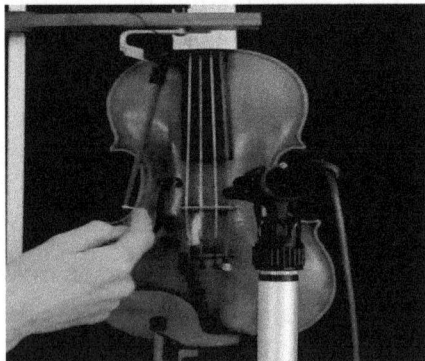

Now the thing about a violin compared with, say, a loudspeaker, is that it's very directional with the frequency above about 800 hertz: it starts getting more and more directional out the front of the instrument. At higher frequencies you get these, you might say, quills of sound in different directions and they move with each semitone.

Gabriel Weinreich was the first to formulate the idea of "directional tone colour"; that a violin's directionality becomes part of what we hear as "violin sound": how it fills space. It's a very particular quality.

He invented a speaker that would reproduce that effect, which is completely lost by a single loudspeaker, because the art of loudspeaker-making is to create something that's as omnidirectional as possible, and as flat as possible in terms of its frequency response.

So it's not a natural quality that you can actually get through a speaker, and most people have only heard violins through speakers. Not much of the world goes to concerts.

HB: I had no understanding of that at all. That alone is shocking to me: the idea of sound filling space in a very different way when you're at a live concert. But it makes complete sense, now that you mention it.

JC: There's been a tremendous amount of research into that lately: how to recreate the musical impressions we have in a concert hall, how to recreate sounds that are satisfying—what we love, what we don't love. Surround sound fills in some spatial information that's lost in stereo that the ear seems to love. At any rate, for me it was a standardized way of measuring the radiated sound of an instrument. I could measure an instrument, then measure another one, and be sure I was measuring the same thing.

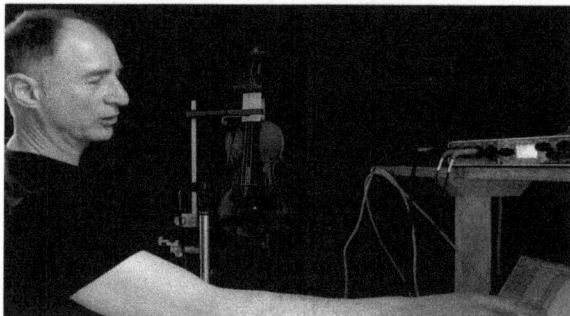

Then I can fold it up, take it somewhere else, and again measure the same thing. One of the problems, of course, is the room: you get room reflections. But I do it with a directional microphone quite close up to the instrument, so the measurement ignores a lot of what's going on in the room. All these things are compromises; you lose some view of the instrument no matter what choice you make.

For the projection experiment in New York, we were able to do the measurements in an anechoic chamber, which is just great. I measured one violin in both that anechoic chamber and right here, in my workshop, and they were very close. There were, of course, some differences, but they were minor.

Questions for Discussion:

1. Were you aware of the concept of "directional tone colour" before you read this chapter? If not, will it influence your choice of seats at an upcoming concert?

2. Might other instruments also exhibit a correlation between frequency and the angle of the projection of their sound? What might this depend on?

3. What do you think the response of professional musicians would be to Joseph's suggestions that they spend some time learning to hear more like recording engineers in order to more objectively qualify the sounds they are producing from their instruments?

IV. Making Modern Violins

In search of a new aesthetic

HB: This all sounds to me like a very interesting and progressive marriage of aesthetics, which I know you care about deeply, with a full embracement of science. My sense is that your goal is not only to objectively determine what the relevant issues are, but also how we can innovate and improve upon our current experiences. As you said, one fundamental aspect is simply basic detection of what's out there: finding out what the sound really is. Because as you were saying, unless we have a clear idea of what's going on, we can't hope to make any real progress.

You've experimented with many things: lighter violins, digital violins and so forth. What was your motivation for doing so, and where are you now?

JC: Well, like all violin makers I started out essentially copying. Whether I thought of it that way or not, I was explicitly copying old violins and trying to look ever more closely at what's going on.

In the mid 90s, after working for about 15 years, I started to feel, as I like to say, like a Civil War re-enactor: fighting battles that were won or lost a long time ago. Why keep re-fighting them? Why do I keep trying to make something that was built two centuries ago when everyone else is moving on? And of course there's this sense that, *Oh, but there is this special something you haven't recaptured yet.* But then again, there's also a certain something about Michelangelo that people haven't captured again. And for good reason.

So the impetus was just restlessness, I guess. I get bored quickly. But I should also say that I'm not particularly fixated on the aesthetic world of the 1700–1800s. When I go to art galleries, I love contemporary and modern art. So there's a real dissonance there. I started asking myself, *Could I make something beautiful that doesn't look like a Strad or a Guarneri? How much of that would overlap with what a musician would like to play?*

This is a fairly conservative world. You wear black and white. You go on stage with your brown or orange violin and you fit into an orchestra. If you want to make a living you've got to more or less conform to that kind of thing.

HB: At least until you get to a high enough level, presumably, when you can do whatever you want. But that probably takes a long time.

JC: Actually, I think now there are opportunities for players to differentiate themselves by doing something new. So much of classical music is fundamentally like museum curating. Think of the Beethoven violin concerto or one of Mozart's, for example. These are absolutely life-changingly beautiful works, but after you've had generations of violinists playing them, soloists start to wonder how they can differentiate themselves.

With the incredibly high level of violin-playing that's going on right now, sooner or later I think players will find that having something distinctly different will be an advantage, especially if they're in a quartet, which is essentially a complete aesthetic unit that doesn't have to fit with an orchestra.

But that's neither here nor there, really. The best direction for innovation is to solve problems, rather than to express aesthetic feelings: try to solve problems with the traditional design.

HB: What kind of problems are you trying to solve?

JC: The violin is often thought of as a perfect design. But if you look closely, it's riddled with unresolved design issues. We makers spend hours making the corners just right. There are whole lectures about how that point is drawn on Guarneri and Strad violins.

HB: Why are the corners there anyway?

JC: For aesthetic reasons. But you give it to the player, she knocks it off, you stick it back on, she knocks it off again, you stick it back on again... At what point is it like a 5 mph bumper? You don't want a fin sticking out in a vulnerable place. Maybe the corners shouldn't be there at all? In our digital violin prototype, the corners are less obtrusive and it still looks kind of like a violin.

Some musicians like it, some don't. But it's a proposed solution to what I think of as a design problem.

But that's the least of it. Violins get damaged in particular ways, and there's no reason we can't design them to resist that damage. For example, there's a post in the violin called the sound post, which is pressed between the top and the back plate. It tends to dig a hole in the top and cause cracks. So years ago, I started putting a little veneer of hardwood in there, which completely stops it from cracking without affecting the sound. Then I started doing the same thing on the other side as well. The strings push the bridge down with about 20 pounds of force into this fragile wood, which gradually works a hole in there. Old violins usually have big dents and cracks around this area. I started putting these little veneers underneath and gluing the bridge to the instrument. It stopped that from happening. These are very mild innovations, but they raised some eyebrows. People said, *"What's that there? How does that affect the sound?"* Well, it really doesn't affect the sound.

Simple things like that are relatively easy to do, although it took me a lot of time to feel comfortable doing them because of an ingrained conservatism. I think it's really the violin makers who are conservative, and the musicians pick that up from us.

Pegs are another issue. Generally, violin makers use friction-fitted wooden pegs, which work pretty well if they're properly adjusted. But if the weather changes, that can affect them. There's no reason why we shouldn't use mechanical pegs. There are some quite good ones coming out now that use planetary gears, which give you a better ratio. In my mind, no one's got it right yet: they're either too heavy or you have to glue them in.

You may also ask, *"Why do we have a scroll here, on the top of the violin?"* Well, because it looks nice; it's just traditional. It has no acoustical effect. You could have an angel's head, a devil's head, or anything you want. It's just an aesthetic thing.

I think the more profound innovations are the ones that try to do something specific to the sound. But you have to say, *"Why would you want to do that? Do players like it?"*

HB: So what are some examples of those sorts of innovations?

JC: Well, the first time Gregg Alf and I had the top off a Stradivari to copy—the client wanted us to put a new bass bar in—I took the top off and weighed it. At that time, people weren't used to weighing violin tops. They measured everything else about the violin, but not the weight of the tops and backs for some reason. You can find books on how the purflings are inlaid, and with what materials, but back in 1985 or 1986 nobody really knew what the weight of the top or back should be, or why you should measure it.

Carleen Hutchins, a researcher and violin maker, suggested that I should get a scale and start measuring. So I did.

I weighed the Strad, and it was something like 54 grams. Mine was 80 grams. That isn't a subtle difference and made me think, *What is the effect of the mass?* And of course, if you step back for a minute, you think, *Why on earth **wouldn't** you measure something crucial like that? Why worry about the subtleties of varnish resins when you could be talking about 30 or 40% differences in mass?*

These ideas preoccupied me for many years: what kind of wood properties would you need to make a lighter weight top? What would the acoustical effects be? The general feedback from engineers and physicists I talked to was that if it was half the weight and everything else was the same, you're going to get twice as much motion for a given force and put out, say, 3 more decibels of sound.

I started making lighter violins and musicians seem to prefer them. Then I thought, *What if you want to make it lighter still? How light can you get?* So I started investigating balsa-graphite composites, graphite-foam composites. We had a fellow called Doug Martin who came to Oberlin. He makes violins out of balsa and they were very interesting.

But my general impression was that if the violin was very light, you would get more sound and a faster response, but you get to the point where it's too loud. The player doesn't really like it. So I started pulling back from that and decided I was going to concentrate on the old sound, to try and get something mellower.

So imagine my surprise in Paris! What the soloists actually seemed to want was simply a great deal of power.

I think the intelligent way to design violins and use materials is to make the kind of sound that the particular player you're working for wants. I know how to make a mellow violin and I know how to make a very powerful one. I don't think you need alternative materials. But there is no reason one shouldn't take the design process further.

For example, when a violinist goes up the neck to play in upper registers, the upper bout gets in the way. Why not get rid of that? So I did get rid of that and made a violin that sounded just like a regular violin. It's been done with violas too. Actually, my first instrument was a cutaway viola with what you might call "a lady shoulder" in this area. But there's still a sense that if you modify anything you're in deep water.

HB: I want to explore this conservatism a little bit. According to my sense of what you're saying, there's this community of violin makers who feel as if they're carrying on the torch of Stradivari, and this notion gets passed on to the players themselves.

But I would also think that some element of a counterculture would exist, maybe not so much in the violin-making world, but certainly among the players. You have so many young people coming up, after all—I guess they're worried about tradition and making their career, but I would imagine there's a certain percentage of them who would be really interested in pushing the boundaries of innovation and being more technically-oriented. We are, after all, living in a high-tech age.

JC: You would think so; and I hope they start to show up. But there hasn't been a call for innovation from players. There's been innovation in string-making which has been very important. In fact, one of the biggest elements in violin history has been the development of strings that work so much better than the old strings did.

HB: Tell me more about that.

JC: Well, most strings that people play now are nylon with metal wound around it. This is done in very sophisticated ways. Strings used to have gut cores, or they would be solid gut, and this made them hard to play because they needed to be relatively thick.

HB: Until when?

JC: The 20th century—in particular, the last half of the 20th century. When I started making violins in the late 70s, the first popular synthetic-based strings were just starting to take over. They are much more stable and easier to play.

HB: When that happened was there push-back from the community?

JC: Oh, you'll always get people saying that they prefer the quality of the old ones. And then there's this whole counter-movement of period instruments, where you get people trying to do the same gut winding that they did at a former time. That's interesting, but there's still this sense of a backward scramble towards the old Italians—as if something has been lost, rather than enjoying all the things that have been gained.

At any rate, no, there hasn't been much pressure for innovation from players as a whole. But there are individual players that are very open-minded.

HB: I wonder if this has anything to do with the general reputation that classical music has: that it's hidebound, conservative, older people's music. It's not generally something you think of as very progressive, forward-thinking, or innovation-friendly. The very expression, "classical music", says it all, I think: it's not just music, it's *old-fashioned* music.

Do you think it's a reasonable fear that unless there is some concerted effort to push the boundaries and make classical music more accessible to people then the entire art form may in fact be imperilled?

JC: I don't think the art form is in peril. Look at what's happening in China or Korea where you have these massive populations that just love classical music. It's huge there. The music education is great there, whereas in North America we're cutting money for public school orchestras. Many of these countries like Korea and Japan have become major players in the classical music world. Whereas the conventional image of a violin prodigy used to be a Jewish boy, now it's an Asian girl.

HB: What do you think is motivating that? Why do you think their appreciation of classical music is so much different than in North America?

JC: Well, you get a population that hasn't heard this body of music, and they hear it and find it very appealing. It's as if we had never heard the Beatles before and then they came. These things just take over. Great music is very appealing.

HB: The Beethoven invasion.

JC: Yes. But I'm not an expert on this; I can only speak from my workshop point of view. What I *can* say is that the Chinese are taking up violin making and they're making very inexpensive instruments that have completely changed the world market. This has made a lot of makers in other parts of the world wonder if they can continue to make a living by making instruments. Of course, that's a different topic altogether.

But to get back to your question about musical innovation, I think there is a lot of innovation in classical music. As far as I can tell, there has been a boom in composition; contemporary "classical" composers are writing music that people love to listen to rather than, perhaps, strictly serial music or atonal music. There is also this phenomenon of important players playing a variety of styles, what might be called world music. Yo-Yo Ma is a great example of that. You also find more use of electric violins, which is another way of tapping into different sounds.

HB: I'd like to get to the question of electric violins, which I know you're also involved in, but before I do I have a few more questions.

Let me ask you about bows for a moment, yet another variable in the production of sound from these sorts of instruments. How much impact does a good bow have?

JC: A surprisingly big impact; and it's not known why. Bow research was hitherto virtually non-existent. There was some good research done, but very little. Recently, it has picked up a bit and there are some very good people involved. But how the bow talks to the violin, how it affects the radiated sound, we don't really know that yet.

HB: What makes a good bow maker as opposed to a bad bow maker?

JC: I don't know much about bows. I've never really learnt. I know a good bow can make the violin easier to play. There's a lot the bow has to do which is very dynamic—good players are moving incredibly quickly. And different bows match different instruments better.

HB: So is it possible to extrapolate from what you've learned in these double-blind experiments, looking past the whole older versus newer aspect and examining which bow fits best with which instrument?

JC: Well, we'll almost certainly begin studying bows in the next double-blinds. François Tourte is the Stradivari of bow makers: he made the best bows ever.

HB: When did he live?

JC: During the late 1700s and early 1800s. He basically invented the modern bow, or at least, he's credited with many of the major innovations.

HB: And that's still considered the apex of bows?

JC: Well, yes. Bows have developed since then, and people prefer different kinds of bows. But he's considered the greatest. His bows are the most expensive. Dominique Peccatte was another famous bow maker—there's a whole hierarchy of bows. Now, if you were to blindfold people and give them different bows, could they tell an old bow from a new one? French bows are like Italian violins; they have that sort of cachet. Can people distinguish a French bow from a German bow or a graphite bow?

After the surprises of the violin tests, I'd be surprised if people can tell as much as they think they can. Although what they clearly *can* tell us is which bow works better for them; and that's really the more interesting question.

I'm almost nervous to start learning anything about bows because it's an utterly fascinating subject. I don't need another thing to flit around with.

HB: Okay, let's move on then. You mentioned briefly that you have also made violas. Have you made any other instruments? Ever thought that it's time for something really big—maybe a double bass in your workshop?

JC: Well, in addition to violas I've also made one cello. I used to call it my first and last. I don't play cello, so I don't have a gut feeling about how it should sound, which makes it harder. But I hope to get back to it. I've got some really good wood and people who are interested, so maybe I'll have a "cello year", where I'll study up on it and maybe take some lessons just to get the feel for it. A self-respecting violin maker should really make some cellos as well.

HB: Oh, really? Is that part of the violin-maker's handbook?

JC: Well, Guarneri, who's arguably the most influential maker now— not Stradivari, people copy Guarneri; that's what players want—he didn't make violas, that we know of. There's one cello that's been attributed to him. But Stradivari made cellos.

HB: How many violas have you made?

JC: I've probably made 50 at least. I usually make a viola or two every year—it just depends on what commissions come along. The viola seems to be ripe for innovation. Viola players seem to be much more open-minded.

HB: Oh, really? There's a different culture there?

JC: Yes. The viola isn't as standardized as the violin. There are a lot of different sizes and there are real ergonomic problems with viola playing, given that it's quite a bit larger than a violin—a cello, of course, is much bigger still, but no one would ever try to stick it under their chin.

So with the viola you've got this intrinsically awkward playing position and it really does make sense to get rid of some of the wood and make it lighter. The violas aren't locked into a single size or a single shape. Working with viola players tends to be a pleasure. They've been much more open-minded in general—so far at least.

HB: I haven't yet asked you about the varnish and that whole aesthetic dimension. That's one of these things you sometimes hear people talking about when they mention the difference between famous old violins and newer ones.

JC: The varnish was considered the most plausible secret for a long time in terms of what made the old Italians what they were. Old Italian varnish was like oriental lacquer: this thing that people didn't know how to make and spent centuries trying to copy—oriental lacquer was this amazing substance that they used to lacquer bowls, cabinets and all sorts of things with, but nobody knew how it was made. It was a very tightly protected secret.

Arguably a lot of European varnishes were trying to imitate this—a friend of mine, Christine Arveil, has made a case for this theory. Anyway, in the end it was discovered to be the sap of a single bush; a very prosaic solution to this big mystery. It was a marvellous substance, but it was toxic to most people. I'm sure there have been interesting books written about that.

Violin varnish was a similar thing. It's beautiful. These old instruments, with their colours worn down... it's a spectacular effect. So there was this search for the Italian varnish. I met a researcher called Geary Baese who was one of the people to put forward a reasonable answer.

He was in Cremona in what must have been the early or mid-80s. It turned out to be just varnish: linseed oil and tree resin. It was a big anticlimax, like so many of these things. For a long time, a lot of the things we loved and didn't understand about these old violins were attributed to the varnish. Then, when people discovered that it was just regular varnish, some people said, *"Oh, but there's this ground underneath it, some crystalline substance."* So there was this little stampede towards pozzolanic cement and silicates. But that's just the history of materials. I think that we now pretty much understand the old varnish.

However, trying to make something look like it's old is something quite different. The thing about old violins is they have a very particular kind of beauty that you don't see anywhere else, because I don't think any other old wooden artifacts were varnished with these transparent, vividly pigmented varnishes that wore off in certain ways. It's an absolutely glorious effect and it's such a perfect metaphor for the sound: this richness, this depth, this light, this sparkle. There are very good reasons why we love them so much.

Of course, as a maker you might wonder, *How can I do that?* But in this case you're not imitating Strad or Guarneri; you're imitating what time did to their work. A lot of makers like to use just a straight coat of red or orange varnish, which is how these old violins initially looked, and see this approach as straightforward and honest compared with "faking" or "antiquing". But there must be other options. All sorts of artworks—some ceramics, some sculptures—have textured surfaces, artificial surfaces, surfaces that refer to the passage of time.

Giacometti is one of my favourite sculptors. You look at one of his pieces and it looks like it could have been dug out of an archaeological site. There's no reason we shouldn't make our violins look a certain way artificially.

What would the new aesthetics be? Maybe something that's been informed by the beauty of the old ones, but is of our time. It's very hard to invent something—I think it will emerge. Will it look more like an iPhone? Will it look very clean, modern, minimalist? Who knows? The main thing is that people have got to start doing it and competing with each other to do it better.

HB: While adapting to the pressures of the modern world, presumably.

JC: Well, we're living in the violin world within the modern world—it's this kind of protected cocoon. A lot of violinists aren't particularly visual people—it's not what they're interested in.

There are some who are acutely aware of the small details, but there are plenty of others who say they don't care what it looks like, they just want it to sound good.

For me it's like saying, "*Do you want your couch to look good or be comfortable? Decide.*" Well, actually, how about **both**? There's no real contradiction there, I think.

I like the electric violin that Alex and I came up with; it's kind of modern and kind of old at the same time.

Questions for Discussion:

1. Why do you think the viola is not as standardized as the violin? What do you think might have been the historical and cultural circumstances involved that led to that situation?

2. To what extent is it possible to objectively assess the societal importance of aesthetic factors in specific times and places?

3. How would you characterize the prevailing aesthetic valuations of our time?

V. The Future Awaits

Digital violins, new books and more

HB: Let's talk in more detail now about your experience with digital violins.

JC: My interest in this came from projects I was working on with Gabriel Weinreich. I built a crude electric violin for an experiment we were doing because Gabbi was interested in how you could make an electric violin sound like a real one using real-time electronics.

We started on that project, but the electronics weren't really up to it back then. Later I was in Cambridge at one of our meetings and Jim Woodhouse and his colleagues were using this new, real-time bit of electronics that would do convolution very quickly. They were using it with what is called the impulse response of a violin, which is kind of an acoustical fingerprint: you could run an electrical violin through this device and it would sound like an acoustic violin.

So Gabbi and I picked that up and it worked out in principle. We made some convincing samples. We had a very good violinist come in and play a few phrases from a concerto on this electric violin and we ran it through the filters and I started laughing because I couldn't tell that it wasn't a regular violin. But at that point we more or less moved on to something else, because that's what you do as a good researcher.

But then it showed up on some online interview, I think, and a young designer from the University of Michigan named Alex Sobolev saw it. He was interested in building electric instruments, so he came over and we started working on a lot of things. And he brought in a friend of his, John Bell, who had just graduated from the University of Michigan and was a sound and electrical engineer. They wanted to start a company that would make these digital violins. John designed the electronics from the ground up and Alex and I worked on the design. We put together this prototype and it's almost market-ready.

The idea is that it's an electric violin that's lightweight. It doesn't feel like a lot of electric violins. In a sense those other electric violins come from the electric guitar world, where you have these big heavy instruments.

HB: How does its mass compare to that of a normal violin?

JC: It's a little bit heavier now, but it will eventually be the same as a normal violin. It also looks, more or less, like a regular violin. With electric violins you often see all sorts of aesthetic statements. We wanted something, at least to start with, that from the audience point of view would read as a violin, rather than as a stick or as a Stratocaster or some other iconic design.

The idea is that a normally-trained violinist could pick this instrument up and it will feel just like a regular violin to play, except the sound will come out of some specially designed speakers. And you'll be able to program in the sound of a particular instrument, whether it's a Strad or a Guarneri, or whatever. You'll be able to change during a concert from one to the other.

HB: And these speakers will be able to reproduce the sound in the way that you described earlier?

JC: That's our end goal. The first is just to have a stereo output that people can plug into a sound system. But yes, we want to make speakers custom-made for this. To my ear, in a lot of concerts that use electronic music, it sounds like all the instruments are just coming out of the sound system. I like the idea of not so much trying to get louder, but having electronically involved instruments, such as this one, with each instrument having its own speaker.

HB: So it would sound fuller somehow?

JC: Yes. You want to be able to localize the sound coming from the individual instruments but have that optimized by some means. We're very interested in having all three components—an instrument, speaker, and electronics—working together in a very flexible, sophisticated way, a way that reflects what the ear wants to hear and what the violinist wants to feel.

At a certain point, I think one could make an interface for designing violin sound in an easy way that would allow violinists to make their own sound. It would be interesting to hear what we would end up with. If you had knobs on a violin that allowed you to change the sound, would everyone end up putting it on the same settings, or would you get more variety? Would you get different settings for different pieces? You can't do that on a regular violin, but with a digital or electronic violin you can really learn about what it is that people want.

What do the players want? That's the big question.

HB: And how have they been received so far?

JC: We took it to *Mondomusica*, an exhibition in New York. It wasn't ready to have players playing it, but there was a very good response. There's an app that Alex has developed to control the violin because it requires an interface. The violin will be controlled through this app and a stompbox. The app will have a library of different instruments you can use to get slightly different sounds.

He also developed this as a tool for displaying instruments. Usually, they're shown as 2-dimensional photographs, but this way you can actually rotate the violin on your tablet device. This got a lot of interest from websites—for violin auctions, for example, it would be an ideal way to showcase the instruments. Violin makers are always rocking the instrument back and forth to see the reflection of the wood; so that's another feature that's been included on this app. Alex did a beautiful job with that.

This is part of the digital violin project, but there are many other possibilities. Old violins are spectacularly beautiful objects but it's difficult to present them in two dimensions and get the feel across.

There are some beautiful books, but they're very expensive: some cost $600 or $700. It's a niche market. But this app will probably cost $0.99 with a subscription for $30 a year, or something like that.

HB: Well, that's wonderful, Joseph—the future looks awfully bright for you and goes off in many different directions, which is presumably exactly the way you'd like it to be.

JC: I enjoy coming at things from different directions. I enjoy researching violins, I enjoy writing about violins. Writing helps me get away from violin making, to have an area where things aren't so regimented. Of course, you end up writing about what you know. I've actually just starting a non-fiction book on the violin world from the inside.

HB: That's going to make you popular.

JC: I've written many articles about the art and science of the violin and profiled many interesting people and researchers. They're just so different from what one might think of as the stereotypes. Players and makers lead very interesting lives that haven't been covered. From the writer's point of view, a great deal has been covered by some really good writers, but I think violin making is fairly open territory.

HB: When are you hoping to finish that?

JC: I've just been talking to an agent and hope to have a book contract soon and go from there.

HB: Well, the very best of luck. I will certainly be on the lookout for it and I'm sure many other people will be as well. Anyway, thanks very much, Joseph. It was wonderful talking to you and visiting your beautiful and stimulating workshop. I really enjoyed it.

JC: It's been a great pleasure.

Questions for Discussion:

1. How might the sorts of digital violins Joseph envisions change the listener's experience at concerts? Do you think such an innovation would be welcomed by most concert-goers?

2. How popular do you think "classical music" will be in 50 years?

3. To what extent do you think the world of stringed instrument makers differs or is similar to that of those who make brass or woodwind instruments?

On Atheists and Bonobos

A conversation with Frans de Waal

Introduction

Aping Morality

A close examination of the combined origins of religion and morality has long been a deeply unsettling experience.

In his dialogue, *Euthyphro*, Plato has Socrates ask: *"Do the gods love the pious because it is pious, or is it pious only because it is loved by the gods?"*

Abstract philosophical mumbo-jumbo? Well, consider the modern theistic version, then, which rippled down like a thunderclap over the centuries, sending monks scurrying to their wine cellars in search of some form of earthly solace:

Does God will the good because it is good? Or is it good merely because God wills it?

If God, who is presumed omnipotent, has even His moral standards constrained by some pre-established measuring stick, then clearly He is not so omnipotent after all: even God, in other words, can't make killing your neighbour the right thing to do.

But then, if God is not constrained by any higher moral guide—if God can decide that killing your neighbour is somehow the right thing to do if His mood strikes—then the difference between good and bad, between moral and immoral, simply becomes arbitrary, a measure of what God saw fit to decide upon on the spur of the moment. That hardly seems terribly satisfying either, to put it very mildly. Hence the need for larger monastic wine cellars.

Nowadays, ensconced in our modern secular worlds where theological disputes can be safely relegated to the quaint and irrelevant, most of us sleep untroubled by these ancient conundrums But we shouldn't. Because they haven't gone away.

Of course, they never did. From Kant to Mill, from Nietzsche to Dostoyevsky, some of the greatest minds of history have butted heads squarely with how we might somehow ground our moral sense in a coherent and meaningful way, with or without God.

Lately, in the United States at least, with the resurgence of gladiatorial-style theatre between strident neo-atheists and unflinching fundamentalists, the public spotlight is once again shining on moral values and religion, but this time largely devoid of any clear understanding of previous insights.

While religious literalists trumpet the Bible as the sole mode of ethical understanding and moral relativists crow that all is permitted, the likes of the astutely self-promotional Sam Harris reveal that science can successfully determine moral principles through a particularly jejune sort of tautological utilitarianism where "human values" are simply those that necessarily lead to "human flourishing".

Into this crowded, if turgid, intellectual landscape, steps Frans de Waal, Emory University's celebrated primatologist.

From a lifetime of empirical research carefully studying the behaviour of chimpanzees and bonobos, De Waal believes that much of what we assume to be uniquely human behaviour is, in fact, shared quite liberally with our fellow primates and other members of the animal kingdom.

> *"If you would say, 'Humans have these tendencies to be moral, but we are the only ones', you would have to explain where such tendencies came from, and you would probably come up with explanations of culture and religion. But biologists, like myself, believe that many of these moral tendencies are much older than our species. It's not something that we invented.*

"Caring for others is very mammalian, so is a sense of reciprocity, a sense of fairness, following rules, and punishing unacceptable behaviour. All of these things can be found in other animals."

In other words, according to De Waal, morality isn't "top-down"—it doesn't come from some pre-set system of rules or some authority figure from on high—it is "bottom-up", something that is naturally contained within us as a species, a value system that arose through a long evolutionary process and which we often strongly share, not coincidentally, with other beings.

To fully appreciate the implications of his claims, De Waal believes, we not only have to reassess our top-down approaches to morality, we also have to fundamentally re-evaluate the hitherto unquestioned uniqueness of our species.

"One of the agendas of my research is to bring humans and animals closer. In a way that means downgrading humans a little bit, because we have a very high opinion of ourselves, and I think we're actually more animal than we tend to think.

But this also means upgrading animals, because I think we sometimes have a very low opinion of animals. And so my goal is a very Darwinian goal, because he was somebody who believed in the continuity between humans and other species; and I strongly believe that as well.

"I think that's a very important message for fields like psychology, philosophy and anthropology, because in these fields people tend to make very sharp distinctions: 'humans are all like this and animals are all like that'."

On the other hand, according to De Waal, there is a clear distinction between humans and most other species on moral issues. While most primates and many other species, routinely exhibit characteristics of what he calls "one-on-one morality", humans are relatively unique when it comes to developing and implementing a larger moral value set throughout a community, what De Waal calls "community concern".

"One-on-one morality is, 'I want to maintain a good relationship with you. I look out for my interests but I also look out for your interests. I will repair the relationship if we get into a fight. I will help if you're down and you will help me—reciprocity.' That's one-on-one morality and there are many good signs of that in primates as well as other animals.

"Community concern, on the other hand, is that I'm not just worried about you and me. I'm also worried about my whole environment and the society in which I live. I worry about whether there is fairness in my society, what kind of rules we follow, whether they're the right rules, and so on. Well, at that level I think we humans develop very differently, we are capable of taking that sort of overview approach, but not many other animals do, in my opinion—I would say that we humans take it much further.

"For example, if I walk through my neighbourhood and see someone breaking into a stranger's house, even though it's not my house and I'm not really interested in what happens there, I will still call the police because I don't want people breaking into houses in my neighbourhood. I have a certain concern about the neighbourhood as a whole because I live in that neighbourhood.

"I'm not saying that community concern is selfless. I think there's a lot of selfish interest involved in me worrying about how my community functions."

Which brings us, rather interestingly, back to religion. After all, humans are the only known species to have (repeatedly) entered into a group conflict—protecting their community—as a direct result of differences in religious ideology.

Does that make us more moral, I wonder? Or less?

The Conversation

I. Denying Our Inner Animal

Cartesian dogs, religious baggage and false dichotomies

HB: My favourite phrase in your book, *The Bonobo and The Atheist*, is, *"I often feel that philosophers should be encouraged to take a pet."* When I was an undergraduate, I was very impressed by Descartes, and I thought, *"Here's this wonderful man, with this incredible intellect, an analytical fellow, and he seems to be able to figure so much out".*

Then I got to the part when he's talking about animals. He was describing dogs and cats as automata, machines that were just processing information and acting in some unthinking, reflex-driven way: clearly they weren't conscious, clearly they didn't have a soul, clearly this and clearly that.

FDW: Didn't he nail a dog to a table or something?

HB: It's possible. But I remember thinking to myself, *"This guy should get a dog."* And if he had a dog, perhaps all of Western philosophy might have changed.

So I just wanted to start off clarifying something: you mention that you often feel like encouraging philosophers to take a pet, but you have actually gone ahead and done so? Have you encouraged philosophers to get a dog? And if so, has it had any effect?

FDW: I'm not sure, but there were of course philosophers like David Hume who was an animal lover. He said there was a continuity between humans and animals, and Darwin was quite influenced by David Hume. But there are many philosophers who come to very strange ideas about animals—as if we're not animals, when we clearly are.

And so they draw this very sharp line: with animals everything is either instinct or conditioning. They hear that from scientists, of course—this comes from psychologists usually—and so they adopt that line and humans are placed in a separate category. Which would mean that in the few million years that we separated ourselves from other animals, we did something very special, miraculous. Which gets us close to religion.

They still have that sort of "religious baggage" with them, which is that something special happened to us.

HB: It's also, in my view, a typical "ivory-tower" type of perspective: they're not only assuming this deeply unique role of humans, but they're so far from experiments, they're so far from actually looking at other species, perhaps even looking at other humans. They're coming up with these grand theories about how the world is from their tower, without actually going out and measuring it.

But of course you and you're colleagues are looking at things very, very differently indeed, going out and directly measuring and interacting with other animals.

Questions for Discussion:

1. What other factors might have driven Descartes to adopt his "automata" view of animals?

2. Do you believe in a "species continuity" between humans and other animals? What does that really mean, in your view?

II. Morality and Evolution

Between chimpanzees and bonobos

HB: So, tell me about bottom-up morality. Let's go right to the heart of things here. What do you mean by that?

FDW: Well, basically we're used to morality being presented as either coming from God or religion, and religion is sort of equal to morality. And after the Enlightenment, the philosophers said, *"No, no, it's not God: it's reasoning and logic that gets us to moral positions."* But that's still a top-down view in the sense that basically it assumes that humans don't know how to behave and someone ought to tell them, or we need to at least provide the arguments for them of how to behave in a particular fashion.

Whereas I believe that we have all the tendencies and possibilities within us to be moral beings, and we actually have a *desire* to be moral beings in order to be accepted in the society in which we live and contribute to. Yes, it is probably necessary to set some boundaries for some people, and that's a top-down approach, but the basic tendencies are already there. And that's the bottom-up view.

The neuroscientists say, for example, that we take pleasure in doing good for others, because they can measure that in our brain—which is, of course, a major basis for moral behaviour since you already have that tendency.

HB: OK. But my understanding is, that this is much more than mere personal speculation. You're grounding these conclusions in your own life's research, you're grounding them in not just humans but looking at other species and being able to see evidence of aspects of what we can call moral behaviour or one-on-one morality.

So, tell me a little bit more about the reasons for your particular beliefs in terms of the primate behaviour as well as other animals.

FDW: Yes. So if it was just humans, it would be a weak argument. If you would say, *"Humans have these tendencies to be moral, but we are the only ones"*, you would have to explain where these ideas came from, and you would probably come up with explanations of culture and religion. But biologists like myself believe that many of these moral tendencies are much older than our species. It's not something that we invented.

Caring for others is very mammalian, so is a sense of reciprocity, a sense of fairness, following rules, and punishing unacceptable behaviour. All of these things can be found in other animals. For example, the reason that the dog is man's best friend is because the dog is able to follow rules, and that's what we like about the dog.

We have all these tendencies already in place, and all we have done in our moral systems is refined them, justified some of them, discussed others, but I think all of that is marginal compared to the basic tendencies.

HB: And you call humans "bipolar apes", because of our evolutionary path: we're almost halfway between the bonobo and the chimpanzee, right? And my understanding is that bonobos and chimpanzees often exhibit quite different forms of moral behaviour from each other, right?

FDW: Yes. Genetically speaking bonobos and chimpanzees are equidistant to us. They have quite different behaviour, and I often feel that we have a little bit of each one in us. It's undeniable that we are a violent species, and that's something we share with the chimpanzee; and actually the reason why we need moral systems is that we are very far from a perfectly nice species. If we were perfectly nice, always friendly, always sharing with everybody, then we wouldn't need a moral system at all. So, we have that chimpanzee part, which can be quite cooperative but also can be very violent.

And then we have all these things in common with the bonobo, which is a sensitive, empathic and highly sexual primate.

HB: You might even say sexually obsessed from what I've read. Or maybe obsessed is too strong a word.

FDW: Obsessed is a bit of a strong word, because they have a lot of sexual contacts but they're very brief, so it's not like they're doing it the whole day. They have a lot of little sexual contact that helps them get through their relationships, basically.

HB: And you talk about how some people who like to imagine that we're primary peace-loving individuals look at the bonobos as their poster boy, as it were, saying things like, *"We must have been descended from bonobos because we're inherently good."*

FDW: Yes, so there's some wishful thinking in that. On the other hand, you have the anthropologists who for years have been arguing that we humans have got to where we are by wiping out everybody else. Basically the idea is that we are an aggressive species and we advance by violence.

That's still a popular view in anthropology and they don't know what to do with the bonobo, because it's equally close to us. While the chimp fits perfectly with that line of thinking, the bonobo doesn't at all. So they try to marginalize the bonobo, and say it's actually not relevant to the story of human evolution. But when genome studies came out on comparing humans to both bonobos and chimpanzees, that tells the story that we're basically equally close to both chimps and bonobos.

HB: So has that actually had an effect in terms of changing the perspective of the anthropological community? Are people accepting this more and more readily?

FDW: No, there's a whole debate going on there. For example, there has been evidence for warfare going on for 10,000–12,000 years, and everything before that time is speculation. There are many anthropologists who speculate that we have an inherent desire to kill and that's why there's warfare going on, but we actually have no evidence for that.

If you look at the Neanderthals who lived something like 40,000 years ago, we've adopted some of their DNA, which means that we did something different from the Neanderthals than just wage war. We may have bred them out of existence basically, which is a bonobo-like strategy, perhaps.

HB: Perhaps not even a strategy, something that just evolved.

FDW: Yes.

HB: So, just to give me a sense of perspective: are you an outlier in this? *"Oh, here's De Waal with his crazy bonobo views?"* Or is this becoming more and more accepted within the community?

FDW: I'm one of the few people that work with both bonobos and chimpanzees. There's a Japanese scientist by the name of Takeshi Furuichi, who works in the field with bonobos and chimpanzees, and he has very similar views to mine: that bonobos are peace-loving and chimpanzees are violent. I've based that mostly on my captive observations, but since I work with both species, I'm interested in the comparison between them.

I wrote a book called *Chimpanzee Politics*, which is all about power struggles and coalitions, and I think human males have a lot in common with the chimpanzee male. And I would never describe the human species as a perfectly peaceful species. I don't think we are.

But I do believe we have to take the bonobo very seriously, and it's very well possible in the longer history of our species we were not nearly as violent as we are now, for example.

HB: So, this would clearly argue for some sort of mixed descendancy: the idea that perhaps we're not exactly in between the two, but that we have aspects of both primates associated with us.

FDW: Yes.

Questions for Discussion:

1. To what extent can we be certain that animals have moral tendencies?

2. How might non-violent bonobos have survived the evolutionary process?

III. The Demise of Veneer Theory

Science discovers human cooperation and empathy

HB: Let me get back to religion a little bit. You have a somewhat different view, it seems to me, than many. You've emphasized this notion of a top-down view of morality: somebody imposes the rules on how we should behave. And the reason for this view, as you've outlined, comes down to what you call the "veneer hypothesis": that we used to believe, not too long ago, that humans were inherently bad, and therefore we had to educate children to make sure they didn't kill each other.

This strikes me as a sort of *Lord of the Flies* notion: that if we were left to our own devices we would wreak havoc and destroy one another. But now the evidence seems to be coming in that things are quite different than that: we've switched 180 degrees to the perspective where it now seems that the prevailing view is that we're inherently good.

How did that actually happen within the anthropological or moral science community?

FDW: Well, it's an old view in the West that goes back to the doctrine of original sin, that we are born bad and if we work very hard we can be good. Freud had that idea in *Civilization and Its Discontents*, which said that civilization serves to keep our basic instincts under control. Morality was looked at in the same way even by biologists in the 70s and 80s; they would write books saying that we're not born to be altruistic, we need to teach ourselves to be altruistic or else it'll never happen.

HB: We need to constrain ourselves as well, presumably.

FDW: Yes. But then around the year 2000 everything changed.

The anthropologists starting doing the ultimatum game all over the world, a game to test fairness, and they found signs of fairness from all peoples all over the world. The economists starting doing games where you could either compete or cooperate and they found that we're actually much more cooperative than we had ever thought. In fact, they started calling us "super cooperators". The neuroscientists found that we get pleasure from giving: if you do a charity test to someone in a brain scanner, the pleasure centers in his brain light up. Primatologists, such as myself, starting saying that animals have empathy and sympathy and have all the tendencies that we see in our moral systems.

So around the year 2000, all of a sudden this veneer theory, as I call it—that we inherently have a lot of bad tendencies but we have this thin layer of morality that keeps it all under control—that basically fell apart. For a while there was still some resistance: people were saying that maybe humans were a very special primate, but now there's a lot of data that shows that the other primates have those same sort of tendencies.

HB: In many of your books you talk about the relationships that you have with chimpanzees. There's a strong emotional attachment, a strong aspect of empathy that you have there. Could you make any sort of blanket statement about the level of empathy between chimpanzees and bonobos? Is it possible to say something with any degree of definiteness there?

FDW: No, we're just at the stage where we're allowed to *talk* about empathy in animals. You mentioned Descartes earlier, and he of course had his followers. For example, the American behaviourists, like B.F. Skinner, had a sort of Cartesian view that animals were basically conditioning machines. And so for the longest time you couldn't even talk about animal emotion, and certainly not animal empathy.

Even in the human literature, empathy was not a subject for discussion. I met one of the pioneers of human empathy studies, a woman called Caroline von Luxembourg, who told me that 30 years ago when she started her studies, she couldn't go to a conference to present on it because people did not consider empathy a serious topic. For them it was classified as something like telepathy, or astrology.

HB: Hold on a minute: did people deny that empathy actually existed?

FDW: It was considered "a woman's topic". It was something that belonged in women's magazines basically, and not in a scientific journal. It was viewed as a sort of "soft science" topic. Now, of course, we consider it a very serious topic, but at the time there was some resistance—even in humans, let alone animals.

HB: And it's now a serious field of research, not only in terms of behaviour but also from a neurophysiological perspective, with all this talk of mirror neurons and the like. How much interaction do you have with neurophysiologists in your field?

FDW: Well, there are some neuroscientists who work in empathy. It's actually quite a small field and I know most people who work there. And mirror neurons, remember, appeared only relatively recently, in the mid 1990s. Nowadays some people talk about mirror neurons as if they explain almost everything but they were discovered quite late in the game.

Questions for Discussion:

1. *Why do you think that the study of empathy was not taken seriously until relatively recently? How do sociological factors influence the type of questions that are deemed to be "scientifically important"?*

2. *Might there be a middle ground between maintaining that humans are "naturally competitive" or "naturally cooperative"? Why do you think scientists are naturally drawn to one extreme view or the other?*

IV. The Roots of Religion

A sociological approach

HB: So I promised to get back to religion, its relation to science and the top-down approach and so forth. Your view seems to be that most standard religious views represent a top-down interpretation of morality and that there are many people of a scientific persuasion that have a top-down interpretation of morality as well.

Intriguingly, rather than looking at religion as something that arises and tells us what to do, you believe that things go in the completely opposite direction. That is to say that you seem to believe that we have this sense of morality inside of us, it is in a way "bred for us" from an evolutionary perspective, and in fact we are pre-programmed to some extent to accept a sense of religious morality. Is that a fair way of describing your position?

FDW: Well, religion, we now think, comes in later. Religion is not the source of morality. Rather, morality originated, let's say, in small group settings during our evolution. And I'm sure before our current religions existed, humans cared about right and wrong, punished bad behaviour, cared about fairness and so on.

But when the societies got too big, like a couple of thousand people or a couple of million people like nowadays, we could not maintain that system because the system is based on me keeping an eye on you, and you keeping an eye on me. We each have our reputations, we try to behave as well as we can, but sometimes we deviate, so everybody keeps everyone else in line.

But, if you have a million people, who's going to do that? There's no way of doing that, and that's when we installed religion which provided, sort of, guidelines and a God who was watching us all the time. And I think that may have helped set up large societies.

HB: OK, but this sounds all very pragmatically utilitarian. If I'm a religious person who has a clear sense in my mind of what's right and wrong, I would probably reply something like, *"OK, you're telling me that people knew the difference between right and wrong earlier on, and that we needed religion to police once we achieved critical mass..."*

FDW: Well, even more than mere policing, because religion also binds people together. Religion has a strong binding function; the influential French sociologist Emile Durkheim, for example, talked about the social functions of religion. The reason why many Americans are religious is because it's a fragmented society where people can move very easily from one city to another and they can use religion to get their social bearings: they can go to the church they belong to in another city and immediately meet another community.

Religion has very important social functions, it's not just policing. And by binding communities together in a ritual-based system of shared belief, you get more commitment to each other. And commitment is a very important part of morality.

HB: Well, I certainly see that religion has a social function, but I had wanted to contrast your views of "bottom-up" morality—that we know what the difference is between right and wrong from our experiences in small groups—and contrast that with someone of a more religious, "top-down" persuasion.

If I were somebody of a religious persuasion, just to play devil's advocate, my view would likely be something like, *"Well, you don't actually know that: you just know what works."* This is what I meant before when I referred to a sense of utilitarianism. In other words, I can imagine an evolutionary process where it's somehow "more effective" evolutionarily to be throwing old women under buses, or what have you, instead of helping them across the street

In other words, just because something is efficacious—in terms of growing your tribe or species or taking over territory or something like that—that it is necessarily linked to what we would normally refer to as something moral.

In other words, if we're getting along—I'm scratching your back and you're scratching mine—we could say that we're bred by evolution to get along because it's helpful to our particular species. But I could imagine a certain circumstance where it would be better for our species if I did something really nasty to you. That wouldn't necessarily imply that I had a clear sense of morality that arises from evolution.

Do you see what I mean?

FDW: Well, the biologist looks at things differently—especially something that is as pervasive as religion. Religion is found in all human communities in the world that we know of, and we look at that with the view that it must serve a function, otherwise why would humans have this desire to form religions?

And the function that I'm exploring at the moment is related to morality and maintaining a moral system, but religion probably serves other functions too. As I've said, there are some general social functions. For example, it gives some solace to people who want to know what happens after death because mortality is a big cloud over our head. Religion probably serves a number of functions that are beneficial.

HB: You also talk about religion being more natural, along the lines by which you've just described, and you contrast it with science.

FDW: Yes. Religion is something that you can teach to very young children, but if you teach science to someone, they will get their PhD at age 33 on average—it's very late. And science came into the game of human evolution very late as well.

There are only a couple of places where science developed in human history, and only very recently—a couple of thousand years ago. Science is also probably the first thing to go: if our society collapses tomorrow, let's say the whole economy collapses along with everything else, religion will stay. I'm convinced of that. But science may go. Science is constantly under attack at the moment, so we need to defend it. I'm a big fan of science, of course, but science is more vulnerable than religion I think.

HB: You used this wonderful metaphor in your writings of science being a cloak that you can wear; and— as you were just saying— easily discard as well.

FDW: You may lose it, yes. Religion, I cannot imagine that we will ever lose it. We may transform it and it may change, but we are very religious in the sense that we very easily believe in the supernatural and follow leaders who tell us how to live and so on.

Questions for Discussion:

1. Do you agree with Frans that science is more vulnerable than religion?

2. Do you think that an "evolutionary view of religion" would be offensive to those who are devout? Why? Would it depend on the religion?

V. Community Concern

Chimpanzee groups and Golden Rules

HB: So, there's this distinction between what we need—emotionally, socially, intellectually—as opposed to a scientific way of thinking which may be beneficial in all sorts of other ways, but is not nearly as strongly linked to our evolutionary needs. Is that a fair assessment?

FDW: Yes, and I also believe that many of the things that we *think* we design by ourselves, by thinking and logic, were actually already there. We're very good at putting our rational labels on things. The sense of justice and the sense of fairness, for instance, we say that that is something that we came up with, but it's based on much more basic tendencies that we can observe in other animals.

For example, in our primate studies we studied the sense of fairness. And we discovered that monkeys have a sense of fairness: that they care about what you get as opposed to what I get, and they get upset if they get less than somebody else. There were philosophers who were quite upset with that result. They had decided in their minds, through very top-down thinking, that this sense of fairness is achieved through reasoning.

HB: Well, you have to get them a pet, you see.

FDW: Yes. Actually, I once met a lady who had a big dog and a small dog, and she said that the small dog insisted on getting as much meat as the big dog. So, the small dog is comparing itself to the big dog, and you cannot explain to him that the big dog needs more meat. So, she tricked him by giving him many small pieces and gave the big dog one big piece, and it worked.

HB: When we examine fairness, there seems to be a difference between individual scales and larger societal scales. In *The Bonobo and the Atheist*, for example, you describe how animals are quite happy to get their reward as long as somebody else doesn't get a better reward, in which case their reward isn't good enough anymore.

But then you make a distinction between this one-on-one sense of morality and a larger community concern, which seems more typically human. So, tell me more about that.

FDW: Yes, well, "one-on-one morality" is: *I want to maintain a good relationship with you. I look out for my interests but I also look out for your interests. I will repair the relationship if we get into a fight. I will help if you're down and you will help me.* That's one-on-one morality, and there are many good signs of that in primates as well as other animals.

Community concern is when I'm not just worried about you and me, I'm worried about my whole environment and society in which I live. And I worry about whether there is fairness in my society, what kind of rules we follow, whether they're the right rules, and so on. Well, at that level I think we humans developed very differently: we are capable of taking that sort of overview approach, but not many other animals do, in my opinion.

HB: Do some others, at least? Do you have some examples?

FDW: There are some examples in chimpanzees. The alpha-male adult chimpanzee will break up fights and he does so impartially: he's not supporting his friends. His best friend may be attacking a female and he's going to defend the female against his best friend.

HB: Because it's in the best interest of the whole group.

FDW: Yes. We've done experiments where you remove alpha males from groups and you get chaos, basically. So this policing and arbitration function of the males is important. Females may fix problems too. So, for example, suppose two males get into a big fight and they're not reconciling. One female will go to one of the males and take him by the arm and move him to the other, and get them to reconcile. Now in order to do these things, that female needs to have some sort of conception of what is good for her society.

I do see the first signs of community concern there, but I would say we humans, we take it much further. For example, if I walk through my neighbourhood and I see someone breaking into a stranger's house, I will call the police because I don't want people breaking into houses in my neighbourhood. So, I have a certain concern about the neighbourhood as a whole because I live in that neighbourhood.

I'm not saying that community concern is selfless: I think there's a lot of selfish interest involved in worrying about how my community functions.

HB: Right, and the argument is that this is a part of us, part of how we look at the world. We could describe it, as many people often do, by invoking a Golden Rule or some top-down level of morality, but in your view that's not really what's going on. You're not calling the cops because a little bell goes off and you think, *Ah, Golden Rule! Do unto others as you would have done unto you*, that's just a description that other top-down people use to describe what's going on.

FDW: Yes. The Golden Rule, the Ten Commandments and other summaries of normative ethics, I think they are imperfect but they do help us a little bit. They give us some guidelines of which direction to go in our moral systems.

HB: But isn't there's also a potentially large conflict here? I mean if the argument is that our morality comes from another level, as you say bottom up, then isn't it potentially confusing for me to be told that, *"No, it's not that we're intrinsically good and have a strong sense of morality from our evolutionary experiences. In fact, we're evil, we're sinful, we have to be told the difference between right and wrong. We have to listen to the experts who can straighten us out by telling us about Commandments and Golden Rules and all of that."*

FDW: Well, The Golden Rule and the Ten Commandments are post hoc summaries. If you want to explain to a child how to live a moral life, for example, you may bring up the Golden Rule. That doesn't mean that as the child grows up she's going to go through the motions of thinking through the Golden Rule each time some difficult decision comes along. But it's a way for us to summarize and justify moral actions. I think that most of our decisions don't depend on those kinds of norms

Questions for Discussion:

1. What, from a biological perspective, might explain why humans take the notion of "community concern" further than most other primates?

2. When the chimpanzee alpha male breaks up a fight on behalf of his community, do you think that he is acting consciously or unconsciously?

VI. Beyond Theatrics

Reconciling science, religion and morality

HB: I'd like to talk a bit about the science vs. religion debates, which you often mention in your writings. My sense is that you're trying to take a middle path, as it were, between the extreme positions on both sides.

On the one hand you have the neo-atheists who are banging the drum on how religion is effectively pernicious and is destroying the rational fabric of our society, and on the other hand you have those of a religious persuasion who are saying something very different: that there's not enough credit given to religion.

My sense is that you're saying, to some extent, *You're both wrong: you're both looking at things from some sort of top-down approach, and you don't realize that we have this morality inside of us.*

Is that a fair summary of your views?

FDW: Well, I think that there are two groups. One is the fundamentalist who goes by the written word of any sacred text such as the Bible. That's not a group that I can work with as a scientist, because they would argue against evolution as a theory.

So that's one extreme. But of all the believers in this country, only one third (I still think that's a lot) are of that persuasion. So there are a lot of believers who are open to science and evolutionary theory to some degree.

Then you have the other group, the neo-atheists, who basically call everyone who disagrees with them stupid. They say, *"We are rational, believers are irrational."*

If you tell all your enemies that they're stupid then they're not going to listen to you. So, I feel like they've worn out their welcome by being so aggressive.

I'm actually more on the side of the atheists, since I myself am a non-believer. But my position is, *Calm down, don't insult everybody, and let's have a discussion about where morality comes from, about the roles of morality and religion in our society.*

HB: So it's more of a tactical issue? It's more of the fact that you support their case in principle, but the way that they're going about their case you find counterproductive?

FDW: Yes, their strategy is very alien to me. I'm from a country where 60% of the people are non-believers and no one blinks an eye if you say you're an atheist. So, I'm used to an atheism that I call apatheism, where you don't particularly care if somebody believes or doesn't believe if God exists or not. This fanaticism that's sprung up is alien to me.

HB: But you've met some of these people, I presume—perhaps even most of them: Richard Dawkins, Sam Harris, Daniel Dennett, and the late Christopher Hitchens of course. Have you had an opportunity to talk to them and tell them what you're telling me now?

FDW: No. Maybe with Dan Dennett, but not with the others. Again, I think it's a strategy that has been counterproductive. If you want to, for example, spread the word about evolutionary theory and why we think we have the evidence to support that theory, then I think it's a bad strategy to trash religion, because you will be offending a majority of people.

HB: Don't you think that realistically—this is my sense of it anyway—they don't really care whether or not the strategy is effective or not. It's more of a self-glorification game.

FDW: It's possible.

HB: Personally, it just strikes me as an act, a sort of circus, that they're all doing in order to increase their book sales. I don't get the sense that there is an open, engaged dialogue that is even attempted!

FDW: I'm interested in an open dialogue and bringing many different viewpoints to the table to discuss. I'm not necessarily against religion, if you're religious that's fine with me, as long as you don't shove it down my throat. To have that open dialogue I think we need to be respectful of others who disagree with us, and of course some of the neo-atheists don't want to be respectful.

HB: We spoke before about your metaphor of science being this cloak that you put on, that it doesn't come naturally to us. You spoke about what Stephen J. Gould said when he was talking about "Darwinian fundamentalists", those who look at every aspect of human nature as necessarily constrained and conditioned through evolution.

It seems to me that we can look at science in exactly this way. Science is something that has taken us an awfully long time to get a grip on, but we can somehow recognize and use that, notwithstanding the fact that it's not as natural as some of these other things. And we will be able to go beyond, or at least sideways, from our evolutionary destiny.

FDW: Yes. We exhibit a lot strange behaviour—we smoke, we masturbate—that is puzzling to a biologist. And there are many genetic characteristics—breast cancer, schizophrenia—that we don't think of as particularly beneficial.

So this whole notion that as soon as people do something on a standard basis, you have to have an evolutionary scenario to explain it, that's not really how I look at things.

And that's what Gould was objecting to: that everything needs to have an evolutionary scenario. We humans are very cultural beings, so there's a lot of stuff in our behaviour that culturally came into being, and we don't necessarily know the reasons.

HB: One more thing before I leave the "science and religion theatre", as I call it. One of the reasons why I'm so sceptical of this, is that it seems to me to be more or less strictly an American phenomenon. You mentioned that you're from The Netherlands, where non-believing is more prevalent, but it's not just The Netherlands. It's actually the United States that's the outlier as far as I can see.

If you look at any Western European country, you don't get this constant invocation of a battle between atheists and religious conservatives. I mean, you go to countries that are traditionally quite religious (Germany in the Protestant tradition, say, France in the Catholic tradition, etc.), and you don't have anywhere near that level of entrenched, theatrical debate between these people as in America. And of course you could go look at various places in Asia and Africa—you could go much further afield.

What is it about this part of the world, the United States, that makes it so receptive to this debate, where you have these constant battles between evolutionary theory and creationism and all of that? Why are they so hung up on this here, you think?

FDW: I don't actually think there was much evolution debate before the 1920s, and of course evolutionary theory is quite a bit older than that. It was somehow manufactured as an issue, basically.

HB: Who manufactured it? How did that happen?

FDW: I don't know how that happened. I don't know how it became an issue, but now we have this radicalization of positions. It's interesting, too, that two of the four principal neo-atheists in this country are not actually Americans but British. And the Brits, of course, have a very long tradition of being anti-Catholic, which goes back to Henry VIII. And maybe that's part of the story.

I don't know how things are perceived in the UK right now, but in The Netherlands, where I'm from, it's not really an issue. People are much more interested in the origin of morality. Whether religion or atheism is right or wrong is not really a big discussion in Holland.

HB: But, I think the two concerns are naturally linked. If I were some-body who had very strong fundamentalist views, then I naturally wouldn't be happy with you saying that morality has evolved from the bottom. I would probably say something like, *"Anything could have evolved from the bottom, whereas I have a clear sense of what is right and wrong because it is spelled out in this book or whatever."*

That's not my position as it happens, but it has a certain consistency, I think.

FDW: In that sense, my position is much easier on the atheist than on the religious person. I give the atheist basically an argument of how you can combine atheism with other things to make it more interesting, because atheism by itself is not an interesting position. It basically says that God does not exist.

But if you combine atheism with humanism or with biology, which is what I'm trying to do, it becomes more interesting.

HB: What do you mean, exactly, by combining it with humanism?

FDW: Well, atheism by itself doesn't propel you in any direction. It only says, *I don't believe in God.* Then the question becomes, *Why lead the good life? Why are you here?* —the existential questions of humanity which humanism tries to resolve. But, humanism actually comes out of religion, and so humanism has a lot of religious elements infused in it.

I think biology can provide many arguments for why we are the way we are, why we would want to be moral beings in a society, and why we would want to have a moral society.

HB: So the idea is to combine atheism not only with humanism *or* biology as you said a moment ago, but perhaps with humanism *and* biology?

FDW: I think so, yes. That's probably the trend that we're going to see, that humanism is going to be combined with biology. Of course, humanism has always put a big emphasis on human nature, but if human nature is also primate nature and mammalian nature, then you get a different basis for that.

Questions for Discussion:

1. Do you agree with Frans that "biology can provide many arguments for why we are the way we are and why we would want to have a moral society"?

*2. What does the word "humanism" actually mean today? Is it possible to **not** be a humanist?*

VII. American Exceptionalism

Speculations on religiosity

HB: So, you shirked my question before about why things are different in the United States. Because it's not just that there is this aggressive debate going on, if you just take a random sample of people on
the street and you ask them whether or not they believe in God, or
where morality comes from, the numbers are completely different
in the United States compared to virtually any other country in the
Western world.

FDW: Also, if you ask people *"Do you believe in evolution?"* Isn't it
something like 40% or so that have doubts about evolution? It's a
very big number.

HB: So I'm going to ask you to speculate on why you think that's so,
but before I do I should also mention this anecdote you wrote about
in one of your books about Stephen J. Gould discussing evolutionary
issues with people in the Vatican who seemed to have no problem
at all with the idea of evolution.

FDW: These were Jesuits, who are a bit more advanced intellectually
than the rest of the Church. They had no problem with evolutionary
theory and they asked where Intelligent Design came from, and why
it was considered by some to be a necessary antidote to evolutionary
theory.

But, to get back to your question, I think Americans are so religious for social reasons. Of course, I'm a primatologist, so I'm not testing this in any way. But I think American society is very fragmented,
very mobile, much more so than what I know in Europe for example.

People move easily from New York to Los Angeles, say, and if you live in such a mobile society then you need some sort of grounding, and it is this grounding that is provided by the churches. If I belong to a particular church, I immediately have a like-minded community wherever I go who will accept me. So, in this way religion has enormous value in American society.

So yes, that naturally translates into large numbers of people declaring. *"I'm religious"* or *"I believe in this or that"*, but I think that the social functions are driving this.

HB: So in your view religion provides this consistent backdrop that is needed because of the mobility of the society, to ground people and give them an appropriate sense of context.

FDW: Yes. And in order to belong to a church, you naturally need to defend the tenets of that church and maybe you become more entrenched in the dogma of the church—maybe that's part of the social bonding process. But I'm just a primatologist, I haven't tested these theories.

HB: I understand. But I'm sitting across from you, so I thought I'd ask. After all, I know you've thought a lot about where our moral sense comes from, you live in the United States and have done so for some time now after having been born and raised in The Netherlands, and you're a keen observer of the rather theatrical public spectacle between militant atheists and religious advocates in this country, so it seemed like a natural thing to ask you.

Questions for Discussion:

1. Do you agree with Howard that the debates between atheists and religious advocates are "mere theatre" or do you think that they serve an important societal function?

2. Is America really so much more religiously inclined than other developed countries? If so, are you convinced by Frans' explanation of why that is so? If not, why do you think that so many people think that it is?

VIII. Testing Morality

Fairness, cooperation, risk-taking and more

HB: I'd like to get back to your research now. You talk specifically about "the moral sciences" and you contrast your view with someone like Sam Harris. In your view, both of you are in the field of the moral sciences, but you don't think he's going about it the right way because he's using a top-down approach of where our morality comes from. But regardless of the differences, it seems to be well understood that both of you are investigating the "science of morality". So my first question is, What *is* the science of morality? What are we talking about here, exactly?

FDW: Well, I do believe that science can help us understand morality: where it comes from, maybe what it is good for, and why we developed it. And actually my work on primate behaviour gives some sort of evolutionary argument on why we developed morality.

But, I would never want to put science in charge of morality and am certainly not advocating that scientists should tell us what the moral decisions are going to be in our society. Science does not have a particularly good history in that regard. I think science has been abused as much as it has been useful for morality.

In addition, that's not the task of science. It's not in the business of deciding how you should live your life and why society should take this or that decision. Science can give you arguments though.

For example, if we're arguing about the death penalty as a moral decision, with science we can muster evidence. We can say, *The death penalty doesn't really help* or *The death penalty is somehow beneficial to society*. So science can really help in these things but still, the decision is up to our society to make.

HB: In other words, science helps by framing the debate in rigorous, analytical, logical terms,

FDW: Yes. And by providing arguments and data, which you need for certain things.

HB: Right. But I'd still like to explore this notion of the "science of morality" a bit, because I have a science background, and for me, there's a relatively clear sense of when something is scientific and when it is not—naive Popperianism, if you will.

Take physics. You guess at what the laws of nature might be, then you test your theory by looking for evidence that will support your claims. Moreover, you try hard to make a claim that can be falsified. And you do the same thing for biology: *I have a theory of the brain. Let's go and test it and see if that's actually the way the brain works.*

But when I think of the science of morality, I'm left lurching, asking myself, *How can I even come up with a claim? Where's the falsifiability in this? What are we actually talking about here?*

FDW: So, let's say that I claim that humans are inherently pro-social, in the sense that we take pleasure in helping others. Well, that's something we can test as well as its being morally relevant. And when we start testing it we're probably going to find the limits, because we're not pro-social with everyone. We're more pro-social within the group and family than outside, so it also shows us the boundaries of our moral systems.

I think in that sense the data can help us understand where these tendencies come from and why we make decisions this or that way. But it's really a jump from there for science to tell us what is moral or not.

HB: OK, so, you can do individual tests on various different claims. Let me ask you some specific questions about that. Has your thinking evolved here, and if so, how?

FDW: In the mid 90s I wrote a book called *Good Natured*, which was on the origins of morality, and I think my thinking on this is now quite a bit clearer. There's also a lot more evidence, actually, on primate behaviour to pro-social and cooperative tendencies, reciprocity, sense of fairness. All of these things we have a lot more data on.

HB: So, to summarize, what did you use to think and how do you think now?

FDW: I used to think that, basically, reciprocity and sympathy were all we had to work with in our moral systems, but now there are many other findings that refine it. For example, this sense of fairness that we test in our primates, that's not something we knew anything about in the 1990s. So that's new.

There's a lot of discussion about altruism and pro-social tendencies that have arisen in the last ten years. When the economists found that humans are quite a bit more pro-social than they had thought—

HB: I'm guessing that you're referring here to the so-called "ultimatum game" you referenced earlier. Could I just stop you there and ask you to describe it in detail?

FDW: Yes, the ultimatum game. Let's say I give you $20 and I say you can split it with your friend but your friend has to accept how you decided to divide it. If he doesn't accept the split then you both lose your money. And so humans under these conditions try to split evenly or 60/40.

HB: Because rationally, as classical economists used to think was the way we acted by "maximizing our utility function" or something like that, I, as the person who decides on what the split is, would give the most to me and the least to you. And you, as the rational agent making the decision, would accept it because it's better get something rather than nothing. But that turned out to be not the way people were acting.

FDW: That's right. And then recently we played exactly the same game—well, with a variation, because we cannot explain the situation to chimpanzees in exactly the same way we can to humans—and the chimps also ended up splitting things evenly.

I used to think that monkeys would care about getting less than somebody else but they wouldn't care about getting more. But now we find that the chimpanzees *do* care about getting more: sometimes they will refuse getting good food if the other one doesn't also get good food. So they also play the ultimatum game the way humans play it.

So now if you ask me what the difference is in the sense of fairness between chimpanzees and humans, I don't know what the difference is. It's not clear to me anymore what the difference is.

HB: And what about between chimpanzees and bonobos? Is there a difference there?

FDW: Bonobos are going to be probably the same as chimps. But Capuchin monkeys, for instance, I think they're very egocentric and they're only going to care about getting less than somebody else.

HB: That makes me think that this sense of fairness, presumably, not only depends on the species but the particular individual, just as it can in human societies. We're all part of the same species but we have different people who are more sensitive and caring than others.

So, what future work do you have planned related to this? What, precisely, are you doing now?

FDW: We're doing tests of cooperation in chimpanzees, where we set up a situation where they can cooperate, where two or three chimps can pull in food together while everyone else is present. That makes things complicated, because I can pull with you, for example, but then the dominant male or perhaps some low ranking female comes in and steals all the food. There's all sorts of competitions, freeloading and stealing.

HB: So, lots of group dynamics come into play.

FDW: Yes. So we want to see how they handle that. The literature claimed that the chimpanzees, as opposed to humans, are not capable of dealing with freeloading because they don't punish whereas humans do punish. But we think that in humans the punishment is exaggerated in the sense that it's not as important as people think it is, while we find that the chimpanzees are perfectly capable of dealing with all this. So we have thousands of cooperative pools and they deal with the freeloaders. Sometimes they punish them but most of the time they sort of avoid them.

HB: Is there any difference between chimpanzees in captivity and the chimpanzees in the wild in terms of how they behave in this regard?

FDW: There is a difference. Chimps in captivity are more intensely social because they're always together, and they have more free time on their hands: they don't need to travel all the time for food. So, there are differences. But the psychology and the intelligence basically stays the same, so they express it in different ways. But generally yes, captivity does affect the behaviour.

It's a bit like comparing humans in our current society: we live in an artificial environment but we still say that we're human. Your psychology, your intelligence is human even though you live in this artificial environment. And the same is true with chimpanzees: it's still chimpanzee psychology and intelligence, but the dynamics of the group are going to be quite different because of the different environment in which they live.

HB: I presume that chimpanzees and bonobos are both endangered, or seriously under threat.

FDW: The situation is really not good at all. The numbers are declining and there's a lot of bushmeat hunting going on in the wild. We used to think, for bonobos, that there were a hundred thousand but

I think the estimates now are more like ten thousand, and that's really not much at all.

HB: Presumably there are various political issues going on as well, because you said that most of them live in the Democratic Republic of the Congo, right?

FDW: Yes. That's a huge country which is as big as Europe. It's a huge country with a lot of forests, and people now have guns which they didn't use to have, so it has become very problematic.

HB: Are there any positive signs through UN cooperation or something like that?

FDW: There are conservationists who, for example, try to teach people how to raise pigs or chickens instead of bushmeat hunting. Some also try to discourage people by fencing off protective areas, but some of the areas are as big as Belgium, so how are you going to keep people out of that?

HB: So, the problem isn't so much the encroachment of civilization, the problem is actually the poachers. Is that correct?

FDW: At the moment that's the big problem. And logging, which brings more people into close proximity with the animals. All of that is problematic.

HB: Is the situation just as bad for bonobos as it is for chimpanzees?

FDW: Yes, I think that all of the great apes are in an enormous amount of trouble at the moment.

HB: Is there any sort of possible way forwards? I mean if somebody wants to help or contribute, is there a fund to support?

FDW: There are local conservation groups and some of them are doing a good job at educating people and protecting certain areas. I'm pretty pessimistic, though. Many of us are pretty pessimistic about what's going to happen there.

HB: But you do much of your research now with primates in captivity. Is that right?

FDW: I work almost entirely in captivity but I have some co-workers who work with bonobos and also with elephants in the field, even though the elephants we work with are under human control, so they're not really wild elephants. I'm more interested in the cognition side of things and for that you need to test individuals.

It's a bit like if you look at a schoolyard with children, you can observe them for as long as you want but you will probably not get a sense of how intelligent children are as representative members of a species until you bring them into a room, put them behind a table, and start doing a test with them. That's where you discover how smart they are; and that's the same with animals.

HB: In terms of your future research, you mentioned studying fairness in group dynamics. Is that your current main area of focus?

FDW: We also work on gambling, risk taking.

HB: How does that work, exactly?

FDW: Suppose I give you two options represented by two stacks. If you take this stack you get very steady rewards, let's say every half a minute you get two rewards. Or you can choose the other stack and you can get very unsteady rewards and some of them are very large and some are very small.

HB: So, what results do you have?

FDW: We're working on that at the moment. There is data but we can't talk about it right now.

HB: Oh, come on—I'm not an academic reviewer or something. I just want to know your current views: are there gambling apes out there?

FDW: We're asking different questions. We're interested in whether males are different than females, because in humans males are greater risk-takers than females.

HB: But you're not going to tell me about any results at the moment. Do you have a hypothesis you can share with me, at least?

FDW: I would expect from these tests that males are more risk-taking in disposition—well, some males at least.

HB: OK. So you're investigating gambling, fairness...

FDW: Fairness, cooperation, cultural transmission—we still do tests on food sharing. For example, I'm more inclined to share food with you if we both obtained the food. Let's say that I obtain a big chunk of food on my own. That should be a different circumstance than if we obtain a big chunk of food together. We've done this sort of thing with monkeys and they do seem to pay attention to that, so now we're doing that with chimpanzees.

Questions for Discussion:

1. If gender differences for fairness, risk-taking and other related factors exist for chimpanzees, what are the implications, if any, for human societies?

2. To what extent do the results of "the ultimatum game" (for humans and also perhaps for animals) impact our understanding of economics?

IX. Reasons for Optimism
Positive behaviour throughout the animal world

HB: You seem like a very optimistic person. It's coming through in your books that you believe fundamentally that human nature is good. One almost gets the sense that you come at this whole issue of the foundations of morality with the *a priori* belief that we're intrinsically good and cooperative. And my guess is that you believe that most other primates also fall into that category. Is that correct?

FDW: Yes, I think the majority of primates are like that: they are friendly and cooperative. But in each group you have a few, in humans as well, that are not. And so in society, how do we handle these people?

There is a pay-off for being not so nice, for being a bully and beating others over the head. This can sometimes get you somewhere. But we need to keep these kind of people under control and that is what we do; and I think that primates do the same thing.

HB: So, it seems to me that there are two possibilities for your worldview. I could say *"Frans had this view even before he became a primatologist; he thought that humans were inherently good, and he looked at the whole of primatology through that filter."* Or, *"Frans had no pre-conceived opinions about this matter but his research studying primates convinced him of this optimistic attitude."* Would it be the first case or the second?

FDW: I think probably the first case. I've always been inclined to be optimistic about humanity in general, and I've always been an animal lover. I got my confirmation for this view when I was put in charge of studying aggressive behaviour, because that was the big topic in the 60s and 70s. Everyone studied violence and aggression, which after WWII was perhaps an obvious choice to study. But in that process I discovered that chimpanzees reconcile after fights.

It's possible that I discovered that because I was inclined to see that kind of thing,

HB: But, the chimps *did* reconcile. It was a real effect.

FDW: Yes, they did. And then we started documenting it. And the first reaction of many of my colleagues was, *"Yes, maybe chimpanzees do that kind of thing, because chimps do a lot of human things, but my animals would never do that."* But now we know that all the primates do it to some degree, one way or another. It really does occur in all sorts of other species as well, including elephants, dolphins, dogs and wolves. It's not limited to chimpanzees.

Questions for Discussion:

1. Do you think that people who describe themselves as "animal lovers" tend to be more optimistic people than those who don't?

2. To what extent is being violent and aggressive compatible with the ability to reconcile?

X. Breaking Down Barriers

Towards species continuity

HB: Two final questions. If I were an omniscient being, what sort of questions would you ask me concerning your work? What sort of questions are you most keen to understand? What are the key items on your research agenda?

FDW: Well, there's one thing that we can never get at, which is what animals feel. I'm a big believer that animals have emotions. We can define aggression or love, but you still never get at the feeling itself. The consciousness and feelings of animals totally escape us, and I think they will forever escape us.

HB: Might neuroscience be able to help us there?

FDW: Neuroscience is already helping to some degree. For example, in the debate about animal emotions, the behaviourist would say, *"Don't talk about animal emotions because we don't know anything about emotions. Emotions are irrelevant."* That's what Skinner would say.

HB: Well, maybe Skinner is irrelevant.

FDW: Indeed, he has become so. But the neuroscientists find, for example, that if you put a war veteran in a brain scanner and you show him images of war, then his amygdala becomes very active. If you stimulate the amygdala in a rat, it adopts fetal postures and becomes very afraid. So, neuroscience shows that the amygdala is apparently involved in fear both in humans and in rats. The neuroscientists are breaking down these taboos by saying that when it comes to things like fear, love, aggression, we well might be able to actually measure these things.

I find it very interesting that the breaking down of the emotion taboo comes from neuroscience. It's possible, then, that we can get at the consciousness of feelings, but I don't know how that's going to happen.

So, if you were all knowing, that's what I would ask you: *what is really going on there?*

HB: Very good. Anything you'd like to add before we wrap up?

FDW: Well, I think we went over a lot of issues. One of the agendas of my research is to bring humans and animals closer. In a way that means downgrading humans a little bit, because we have a very high opinion of ourselves, and I think we're actually more animal than we tend to think.

But this also means upgrading animals, because I think we sometimes have a very low opinion of animals. And so my goal is a very Darwinian one, because he was somebody who believed in the continuity between humans and other species, and I strongly believe that also.

I think that's a very important message for fields like psychology, philosophy, and anthropology, because in these fields people tend to make very sharp distinctions: *humans are like this and animals are like that.*

HB: Are we moving in the right direction, you think?

FDW: Oh yes. Over the last 25 years it's incredible what has changed. Twenty-five years ago Ed Wilson was talking about social biology, the biology of human behaviour, and he got so much flack for it. He was called a fascist because he wanted to discuss the biology of human behaviour. That has completely changed now. Now when I talk to audiences about comparing humans and animals, there's is none of the hostility I would have received 20 years ago.

HB: And when you give public talks, are they better received in some places than they are in others?

FDW: They're very well received here in the US. You might think that, with all that we said before about evolution and religion in this country, that people would be very upset, but that's not been my experience. Of course, I may be self-selecting a certain type of audience.

HB: I suspect so.

FDW: But on the other hand, if I'm invited by a zoological society to give a talk at a zoo, that typically represents a cross-section of a city. They're not just atheists—there are believers among them I'm sure—and the talks are generally very well received.

Whatever people's feelings are about evolutionary theory, that connection between human and animal behaviour is very basic. They can recognize that, and they are willing to recognize that.

HB: Especially pet-owners—once again, I think the central message here is that everybody should go out and get a pet. Especially philosophers. But not just philosophers.

Well, thank you very much, Frans, it was a pleasure talking to you.

FDW: Good talking to you.

Questions for Discussion:

1. Has this conversation changed how you feel about the origins of our moral sentiments?

2. Has this conversation influenced your view on the relationship between humans and other animals?

Continuing the Conversation

For a more detailed appreciation of Frans' views, readers are directed to his many books, such as, *The Bonobo and the Atheist: In Search of Humanism Among the Primates, Mama's Last Hug: Animal Emotions and What They Tell Us About Ourselves, Are We Smart Enough to Know How Smart Animals Are? The Age of Empathy: Nature's Lessons for a Kinder Society, Chimpanzee Politics: Power and Sex among Apes.*

Embracing the Anthropocene

Managing Human Impact

A conversation with Mark Maslin

Introduction

On Being A Superpower

Contemporary physicists like to invoke something called "the Copernican principle", named after the 16th-century Polish astronomer who famously upended our self-important and largely unquestioned assumption that everything revolved around us. The modern version goes considerably further than anything Copernicus had in mind: rather than replacing the Earth with the sun at the privileged position in the centre of things, today's cosmologists believe that there is, in fact, no "centre" at all—if things appear uniformly spread out from our vantage point then it's safe to conclude that it looks that way from every vantage point because there's nothing particularly special about ours. Unsurprisingly, perhaps, this general outlook is also sometimes called "the mediocrity principle", which urges people to assume that, unless we have some clear reason to suspect otherwise, any given scenario is as likely as another; and so there is nothing terribly unique about our own situation.

As a way of deliberately ridding ourselves of the many human-centred biases that distort our perceptions there is much to recommend this approach, and it's generally regarded that this steady determination to take humans out of the picture is directly responsible for the remarkable scientific and technological progress we've enjoyed over the past few hundred years.

But now might well be the right time to consider putting humans back in again, with many opting to describe our current geological epoch as "the Anthropocene" given the highly significant impact of humans on the planet.

As Mark Maslin puts it:

> *"For the last 500 years, the fairly constant scientific message has been that individual humans are tiny and insignificant: it's a big, dangerous, nasty universe out there, and we're just one small part of nature that has evolved.*

> *"However, the Anthropocene says, Well, hang on; we're actually the most important thing on this planet, which is the only planet where we know life exists in the whole universe; and we control its destiny."*

Mark, a geographer at University College London whose research spans climate change, human evolution, and global development, is clearly not saying this merely to improve our collective self-esteem, but rather to impress upon us the frightening power that we, as a species, now possess.

> *"This prompts us to ask, 'OK, then: what sort of future do we want for this planet?'*

> *"People are now starting to talk about 'bad Anthropocene' and 'good Anthropocene'—contemplating different possible futures that we collectively can imagine for the future: whether as custodians of the planet we make sure that we keep everything as nice as possible for all of us—which means that we actually look after all of us and the planet—or do we just have business as usual and keep going as we are because, well, that's just what we do. For me, then, the big thing is that we've now hit that choice of which direction to go."*

But in order to move forward to do that, it's worth recognizing that many need to be convinced that the concept of "the Anthropocene" makes sense at all.

As it happens, Mark is a specialist on "the Anthropocene", not only through his professional ability to fully appreciate the detailed scientific evidence for how the Earth's climate has changed over decades, centuries and millennia, but also because, as we've seen above, he recognizes the potential impact of that term: if "the Anthropocene" is going to really grab our attention, in other words, it should mean more than just "humans are currently influencing things".

Which is why he has spent considerable time and effort in putting forward a clear position of when we should say that such an epoch actually began.

> *"There are a number of ages that have been suggested, and there's even a discussion whether you need an actual proper geological marker in time, or whether you can just make up an age.*

> *"Simon Lewis and I wrote this big review paper in Nature about it. We argue that you have to stick to the geology, because if you don't stick to the geological definitions then you're basically just playing politics. So you have to go back to the fundamental science.*

> *"We found in reviewing everything, all the different possible dates, that there were three."*

Those three are:

1. About 5,000 years ago when the cumulative impact of agriculture began to be noted in the CO_2 and methane geological record.

2. Around 1610.

3. In the 1950s, during the so-called "Great Acceleration" when significantly increased levels of CO_2 began being injected into the atmosphere.

Mark and his colleagues opt for 1610. Why 1610?

> *"What you find is that there is a dip in the CO_2 record about 1610. And the reason for that is due to the European colonization of the Americas. By doing that, what they did was start to exchange organisms from the new world and the old world.*

> *"So you have this huge swapping of organisms, which is irreversible. Unfortunately, we also bought over smallpox, measles and typhus to*

the Americas. And present estimates suggest that 50 million people died in the Americas.

"And the result of that human catastrophe was that the rainforest and the savanna grew back. And as it grew back, it took CO_2 out of the atmosphere, so much so that you can see in the ice core record that CO_2 globally dropped.

"That is why we argue that this was 'the start' because if we all disappear now, that exchange of creatures is permanent in the record."

And now Mark's point becomes clear: the Anthropocene should properly begin not only when we can clearly establish human impacts on the planet, but when such impacts have irrevocably changed the course of Earth's biological and geographical trajectory. That is what it means to be, as he urges us to recognize, "a geological superpower".

"The impacts of being a geological superpower certain include climate change (and how to address it), but also many other extremely important issues that a single-minded focus on climate change might have missed.

"For me, this is why even the discussion of the Anthropocene is really important.

"Climate scientists have become louder and much more insistent over the last 25 years about climate change because they felt that nobody's listening. So they keep pushing and pushing and pushing.

"Unfortunately, what that's done is to allow other things that are happening due to human activity to be missed: the huge loss of biodiversity, the overwhelming amount of deforestation, the amount of changes in natural habitats, our alteration of even the nitrogen cycle—because, at the moment, half the nitrogen fixed from the atmosphere is fixed by us to make fertilizers, to produce food—all of these things, which show that our impact on the planet is much wider than only changing the climate."

So the good news is that we're clearly not insignificant—we are, in fact, nothing less than a "geological superpower". But with that power comes a deep responsibility. And it's high time we started taking it very seriously.

The Conversation

I. Becoming A Geographer

A serendipitous journey

HB: I'd like to start with your background. Have you always been interested in science from a young age, or was that something that came later?

MM: I'm of a certain generation whereby we only had three channels on TV. Of course one of them was BBC One; and in the mid '70s, the first *Life On Earth* series by David Attenborough came out. So at the age of about 8 years old, I was taking this on board, this incredible flagship program showing science at its highest level.

We also had a science program called *Horizon*, which still runs today. But in the '70s, it was pitched at above university level; and so as an eight-year-old, I would devour these hour-long impenetrable science programs and not understand them.

I also have a brother, and my parents had a problem: what do you do with two boys who are fractious, who fight, on a wet Sunday afternoon in London? So they used to get in their clapped-out Cortina and drive us down to the Science Museum, or the Natural History Museum, and literally throw us in there.

So my whole youth was based on the idea of understanding science through these beautiful and amazing museums, as well as *Life On Earth* and *Horizon*. And I think it was quite early on—it was definitely before I left junior school at about the age of 11—I had decided I was going to be a scientist. I wasn't quite sure which one, though.

HB: What about your brother? Was he influenced by them the same way?

MM: No, interestingly enough, my brother went down a slightly different route and has become an accountant. So the same influences clearly don't push people in the same direction.

HB: It's not a one-to-one map.

MM: No. Otherwise, we'd all be scientists after having watched David Attenborough.

HB: Or we'd all be accountants, God forbid.

MM: Yes.

HB: And you said you weren't entirely certain which type of scientist you wanted to be at the time. Did they all hold equal allure for you? I know you went into physical geography later, which I hope to get to later. Were you particularly fascinated by the evolution of the earth? Rocks and volcanoes? How the solar system formed? Big Bang cosmology?

MM: I had two areas that I was interested in. The first was of course, space because in the mid-to-late '70s, we were all still filled with this buoyant idea of the Space Race: we were going to go and live on Mars and the moon in only a couple of decades. The idea was that I would be having space trips as an adult—this was in the psyche of my generation.

And then, there was a general fascination with how everything fitted together, because I'm a systems person: I love trying to work out how things fit together. So from school, I was starting to get this idea that the atmosphere is attached to the land, which is attached to the oceans and so forth—and it was that sort of system that started to really engage me.

That's where my studies started to go and why I went off to university to do both geography and geology at the same time. But at the back of my mind, there was always this sense of, *Well, hang on, we're all going to be in space*. That was my default: *If we can all be astronauts, then I'm going to be an astronaut.* Unfortunately, it didn't pan out like that.

HB: Was there ever a moment when you finally recognized that you weren't going to be an astronaut, or was it more simply a gradual deepening of interest in geography and geology?

MM: I think it was when the space shuttle program closed down that suddenly you could almost hear the door slamming shut. And even though I'd gone in a completely different direction, there was a collective sigh around the science world, with people saying, "*Oh. Does that mean we're not...?*"

Because again, science and NASA and all of that had been pushing the frontiers of science in one direction, so we felt quite right to push it in other directions at the same time. And I think that was quite a shock to people. I don't think they necessarily saw it at the time, but again, we still hanker after that excitement of the space race, to actually get out there, to push those frontiers. Whether it happens to be into space or it happens to be into the deep ocean, wherever we haven't been, we really want to go. That's what makes us humans.

HB: You mentioned your parents dragging their fractious children to science museums.

MM: Yes.

HB: Now, however effective such a strategy might have been, I'm guessing that there's probably more to it than that, given that there are all sorts of other things one could be doing to calm one's children down. Was there a clear endorsement of scientific activity in itself? Did you talk about science around the dinner table? Was there any of that sort of thing?

MM: I think the key thing that my parents did was just open up the way. They were always very supportive, but they had no concept of what I was doing. I was the first person in my family to go to university, so I was plowing a new field, completely against the grain of the whole of my previous family.

They wouldn't stop me, but they always said, *"Well, look, if you don't want to do that, there's always other options."* So they were always very supportive and ready to catch me. And they'd keep looking at me at each stage of the academic ladder, watching me as I navigated each stage.

In my family it didn't actually matter what the children do, my parents just wanted to be as supportive as they could be. They don't have any expectations.

HB: Clearly: your brother's an accountant, after all.

MM: Well, yes. So it's an interesting thing because my parents didn't set any expectations, but it also means that I can't achieve anything because there's no expectations to compare myself with.

HB: Did they also enjoy these television shows that you were talking about? Were they also avid followers of the Space Race, or of the latest, greatest thing that was happening in science? Or was there not so much of that type of enthusiasm?

MM: Well, when you only have three channels and your parents are in control of the TV, the viewing was obviously controlled. So I remember watching *Horizon*, the science program, with my father, because he was always into understanding science. On a Sunday night, we used to watch *The Money Programme* because my father was really interested in how to actually manage our limited finances.

My family is more BBC Two than BBC One, which meant that soap operas and things like that just didn't really come into my family's psyche.

And they were also very good about trying to give my brother and I tasters. So when they could afford it, we went to classical concerts or would go to the theatre. They'd give us little tasters of things out there, which was as much as they could do with their limited finances, but it was just enough to give my brother and I a different view of the world and allow us to see that there were things beyond what we were doing in order to help us make our own way in the world.

HB: Did you have a particularly influential teacher when you were in high school, that influenced you?

MM: There were a number of teachers that have inspired me—interestingly, not always in science. One of my best teachers was in history. His way of teaching 20th-century history was just absolutely phenomenal, and has always allowed me to keep that interest in human history and human development.

HB: What did he do in particular? What was so remarkable about the way he taught the subject?

MM: Again, it's sometimes about the structure of how you tell a story, but it's also about narrative. This is something that I try to bring into my teaching: the idea that each lecture is a story, because people, from the very early days or our evolution have sat around fires, telling stories. And that is an incredibly powerful way of teaching and bringing the new generation into new ideas. It's not just to say, "*Here are the facts,*" it's to link them with stories: humanize them by making sure they understand who's actually come up with the ideas, why they had to battle to get them. So I think that's really important; and that was a product not only of my university days, but also my school days.

I can remember at the beginning of my undergraduate degree at Bristol, in the first year, being taught about human evolution. That lecturer was probably the worst lecturer on the planet. They were the dullest lectures, they were so dry. But the actual underlying science was so fascinating that I went and toddled off and started reading up on my own. My curiosity is one thing that's just driven me on and on and on, because I just want to understand. And interestingly enough, over the last 15 years one of my major scientific fields has been looking at early human evolution in Africa.

HB: I was wondering how that came about.

MM: It started in year one of university with some of the worst lectures I've ever attended.

HB: Was this an elective? I wouldn't think that would be a mandatory course for a geography major.

MM: Well, geography in the United Kingdom covers a huge range. In my own department at UCL, we will do things about deep-sea sediments, long-term climate change, the ice ages, all the way through to hydrology, geomorphology, all the classical physical geography, including weather.

And then we get into the social side: economic development, all the way through how different cultures view nature. We even have one person, James Kneale, who studies science fiction novels, something I find incredibly fascinating. He looks at how science fiction has influenced scientists, and how science has influenced science fiction.

Because I have to say most scientists will guiltily put their hands up and admit that we read a lot of science fiction when we were kids. And again, that gives you that vision of what the future could look like. And of course, most visions of the future are driven by new exciting science.

HB: And then after your undergraduate degree you go off to Cambridge to do graduate work in paleoceanography, right?

MM: Correct. One of the things I say to my students is that you never know which direction your career is going to go in. And sometimes, actually, it's the rejections that define your career. So in my third year of undergraduate studies at Bristol, I toddle off to Cambridge to be interviewed for a PhD in the geography department on eskers. Eskers are S-shaped hills that you find formed under great ice sheets. I would have to go to Finland, drive around, measure them, and so forth.

Now, two of us were interviewed. One had already graduated from that department with a first, had a career already in computing—which in those days was something quite special, actually having touched a computer.

And of course at the end of the afternoon, that person got the PhD. And one of the people, who's still at Cambridge, grabbed me and said, "*I'm not happy with this. You should have got that PhD.*"

He dragged me across the road, which happens to be called Downing Street. So I crossed Downing Street, was taken into a rabbit warren and introduced to Nick Shackleton. And he then sat down for about 10 or 15 minutes, flicked through my third-year dissertation and asked me lots of questions. I then had afternoon tea, because that always happened in the Godwin Lab. And then he showed me the door and pointed me towards the direction to the station, telling me, "*Phone me tomorrow.*"

So I phoned him the next day. He said, "*Oh, well, if you get the right degree, then yes, you can have my PhD.*" So that's how I got the PhD with Sir Professor Nick Shackleton, FRS (God).

It actually took me until about the second year of my PhD to realize how famous and important my supervisor was. So again, I try to teach my own PhD students that sometimes failures and rejections just open new doors. My career was pure chance: I was at the right place at the right time and actually got turned down for what would have been a really bloody awful PhD.

Questions for Discussion:

1. How many current scientists do you think were strongly influenced by television programs when they were children? Would they have become scientists anyway, had they not watched those shows? Are we currently producing enough good-quality programs to inspire the next generation?

2. To what extent do you think a manned space mission to Mars would inspire scientists of the future? Would it have a significantly greater effect than unmanned probes?

II. The Anthropocene

Exploring three starting dates

HB: I sometimes wonder about how somebody who has the sort of background that you have, who's clearly interested in understanding the science of the earth, examining what's happening in the oceans and comparing the archaeological record with present-day trends, analyzing the data through various models and so forth, finds himself spending an inordinate amount of time writing op-eds, appearing on stage in debates, and trying hard to disabuse people of their false opinions about climate change. That's a bit of a shift. Do you ever ask yourself, *"How did I get here exactly? I used to be really interested in all this quantitative science and I wind up with all this political stuff"*?

MM: I do wonder how my research agenda has got so wide; and I have to say, people do occasionally raise eyebrows. So for example, last week, my PhD student and I have a "Commentary" in *Nature*, which is on global clean-energy negotiations and how different countries are trying to set their limits, which is a considerable distance away from how the ice ages started.

But I think it's partly related to starting off as a geographer, which many people see as a bit of an insult. So, if you mention that you're a geographer to a physicist, they'll just go, *"That's not even a scientist."*

HB: Sure. But those guys are terrible. Everybody knows that.

MM: Yeah, I can say that to you.

However, when you look at any natural system, the first thing you learn as a geographer is that it isn't a natural system because you have humans. The human influence on every single system is always there; and therefore, even if you try to scientifically focus on the natural processes, you then have to make a special effort to also take into account the human processes. So that's always been there.

In addition, while science is great and I love science, I'm always aware of the fact that science then has to interact with society: my salary is paid for by the UK taxpayer. So at some point, I need to say *why* I'm doing the science that I'm doing and what relevance it has.

And I think the key thing—and this is something that I try to be very careful of—is that I stick to my knowledge base. So, for example, if I'm talking about climate change, I will stick to the parts of climate that I know. I also make it very clear to people when I'm talking about what the science tells us as opposed to what my opinion is; and I try to make sure that people know that there are two different things going on. Because, of course, it is very easy as a scientist to get sucked into that political maelstrom and find yourself putting your own views out there as supported by science.

So again, I'm very clear about the difference between *me* and *the science*. And I think if you do that it just adds to the variety of your experiences. I think that many scientists have become too focused. If we go back to the natural scientist of the Victorian period, where you had knowledge of many areas and how they intersected, as well as how it interacted with society, I think those individuals are much more interesting and exciting than those who are just doing one thing for 10, 20, 30 years.

And all of this work about the politics, about humans and about natural systems has actually come to fruition in the last couple of years because scientists are now asking, *Are we in the "Anthropocene"?* The "Anthropocene" is this wonderful term that's been coined that suggests that influences of humans are so large that perhaps we can be perceived as a geological superpower. That is, we're now influencing the planet as much as say, plate tectonics or a meteorite impact; and therefore, we're having a planetary-scale effect, both on the environment, but also evolution—because we're now controlling evolution through extinctions and through new species that we're creating in the labs.

So we've suddenly become this power. And to actually recognize that, perhaps we should look back and say, "*Well, hang on. When did we start a new geological period?*" Or it would be called an epoch.

And that debate is really interesting, because it brings together all the natural systems, all our impacts, and then you put the sociological dimensions on top of it to ask, *When did we actually become this super-creature, actually affecting the whole planet?*

So bizarrely enough, it seems that all my background has slowly been moving towards this pinnacle of now being able to discuss and actually argue about when the Anthropocene started.

HB: So when do you think the Anthropocene started?

MM: Well, I try to be even-handed with the Anthropocene debate, unlike some of my colleagues. There are a number of ages that have been suggested, and there's even a discussion whether you need an actual proper geological marker in time, or whether you can just make up an age.

Simon Lewis and I wrote this big review paper in Nature about it ("Defining the Anthropocene", March 11, 2015). We argue that you have to stick to the geology, because if you don't stick to the geological definitions then you're basically just playing politics. So you have to go back to the fundamental science.

We found in reviewing everything, all the different possible dates, that there were three. One's about 5,000 years ago, when early farmers began to deforest large areas of the world and started wet rice paddy fields—at that time there is the first increase in CO_2 and methane, that we find in the ice core records that influences global climate.

That may have even actually delayed the next ice age, that small change there due to the cumulative impact of agriculture up until 5,000 years ago. It's only about 40 parts per million, but it may have just taken us up over the CO_2 level that would have dropped into the next ice age.

Now CO_2 and methane continue to rise fairly linearly through a long time period thereafter, from about 5,000 years ago to about

1,000 years ago. So that does that. However, there's nothing else that happens—there's no other major changes.

The next date that Simon Lewis and I really like, and are pushing, is 1610, because what you find is that there is a dip in the CO_2 record about 1610. And the reason for that is due to the European colonization of the Americas. By doing that, what they did was start to exchange organisms from the New World and the Old World.

So you have this huge swapping of organisms, which is irreversible. Unfortunately, we also bought over smallpox, measles and typhus to the Americas. And present estimates suggest that 50 million people died in the Americas.

HB: 50 million? Wow, I knew many people died, but I certainly hadn't appreciated it was that much.

MM: That's our present understanding—50 million indigenous people. Because what we've discovered by looking at the archaeological records properly is that actually, we're very successful as people, and we're very successful as farmers. So we now know that in the past, there were a lot more of us than we were expecting—so yes, 50 million.

And the result of that human catastrophe was that the rainforest and the savanna grew back. And as it grew back, it took CO_2 out of the atmosphere, so much so that you can see in the ice core record that CO_2 globally dropped. And then as population recovered, it recovered and we moved towards the Industrial Revolution.

That is why we argue that this was "the start" because if we all disappear now, that exchange of creatures is permanent in the record.

I'll give you an example. If you go to North America, the majority of earthworms in North America are European. So we can't reverse that. And the reason being is because European earthworms have this little trick, which is they go through the soil and then they go up, grab the leaf litter and pull it down. North American earthworms never had this trick: they just wait for it to decay, to get into the soil, and then they'll eat it. So of course, the Europeans began to completely outcompete the North American native worms, and that could never be reversed.

At any rate, our position is that 1610 is the start date of the Anthropocene for the reasons I just gave; and that was our novel contribution to the debate.

And then the other date that a lot of people really like is what's called "the Great Acceleration". From the 1950s onwards, everything starts to change: from 1950 to today, we've doubled the world's population, we've increased the amount of deforestation. It all started to become exponential in the 1950s.

And there's a very clear spike that you can use, a geological spike, which is the C_{14} record. Up until 1963, we were testing nuclear bombs above ground. Oops. Of course, you have all that radioactive material kicking around in the atmosphere, and that's picked up particularly in radiocarbon. So you have these increase from 1945 with the first bomb, all the way up to '63. And then in '63, there was the Partial Test Ban Treaty, which stopped above-ground bomb tests, and then it's dropped off since then.

So globally, it peaks in 1964 because it takes about a year for it to mix, and it gives you a lovely spike, right in the middle of the 1960s, which could be used as your stratigraphic marker. Because again, huge amounts of things changed during the '50s and '60s.

In the '60s and '70s, CO_2 starts to ramp up massively. Then, in the '80s and '90s sea temperatures and air temperatures start to increase rapidly as well. So that's another possible date to choose: those are the three possible starts for the Anthropocene.

HB: Okay. So I'm going to ask you to elaborate on why you opt for the one you do—the 1610 date—but before I do, there's one question that came to mind as you were speaking about the effect of above-ground nuclear weapons testing.

My understanding is they tended to do this more in the South Pacific and other places. Now, I understand that there are convection currents and that if you do something major like that somewhere in the world, it will affect other places in the world, but to what extent are those geological markers localized so that you see a certain phenomenon much more in one place rather than another?

MM: Well, there are tree rings. So, what you do is you get beautiful tree rings from the '50s, '60s and '70s, and we've got them from many different places: the Northern Hemisphere, the tropics and elsewhere—a friend of mine has just got a beautiful record from Tasmania. And then you measure the radiocarbon in the tree wood. What you find is that there is a peak in one place and then a year or so later there is another peak elsewhere, because the atmosphere mixes so quickly you only have a year's delay. So, you have this beautiful one year peak, which then ties in.

HB: When I started thinking about tree rings in preparation for this conversation, once again I started to appreciate how much I didn't actually know.

I get the idea that you can test for trace amounts of something—oxygen isotopes or carbon or whatever. And I can imagine that there are some trees in Canada, for example, that have fairly well-defined growing seasons so you might get a ring structure where you can clearly tell when the tree was growing and when it wasn't. But if you're looking at flora in subtropical zones, they might not even have rings. Is that right? I mean, does it matter where you're looking?

MM: Oh, I have to say I'm not an expert on tree rings.

HB: Well, you're definitely the expert in the room.

MM: Well, I suppose—I do make sure that I cover this in my master's courses. So yes: trees in different climate zones have different structures. In the Northern Hemisphere, where most of the work's been done, we have tree ring chronology going back 14,000 years or so, linking tree to tree from the present day all the way back to 14,000 years—a beautiful chronology that we get by matching the thickness of the rings.

You can overlay each of the records, and this has been done in great detail for all of Northern Europe. And what's interesting is, of course, you do get two rings, which is of course, the winter and the summer. Now, as soon as you get into the Mediterranean, where I take my third-year students on field work and they do tree rings, you also get two rings, but that's because of the wet and the dry season, which is summer and autumn whereby it gets too cold in the winter for them to grow. So it's different.

And then when you get into the tropics, what you may have is rings that correspond to wet and dry seasons. So, you have to know your tree, you have to know your species and each species has different ways of growing.

This is why people spend their whole lives just working on tree rings, to be able to understand the tree and the associated ring structure. As I teach the students, you have to look at lots and lots of trees, because some trees will grow crooked and the rings will be twisted or you get wind, resulting in trees with narrower rings on one side than the other side. So it takes a lot of work. That's why I'm not a dendrochronologist.

HB: Very good—thanks for that. Now getting back to the Anthropocene...you mentioned that there are three objective geological markers, three possible dates to choose for the beginning of this period: the cumulative effects of human agriculture that produced an increase in the CO_2 record about 5,000 years ago; a decrease in CO_2 around 1610 due to the interchange of species and a consequentially massive decline of indigenous populations in North America that is associated with the regeneration of the rainforests and the savanna, and then the "Great Acceleration" in the 1950s resulting in a strong increase in CO_2 levels. And you and your co-author of this Nature paper opt for 1610. Perhaps you can embellish a bit on that.

MM: Yes, Simon Lewis and I are strong advocates of 1610 for three reasons.

First, the exchange of plants and animals—horses, cows, cassava, maize to Africa—is a permanent signal that we can never undo.

Whatever happens in the future, we have changed evolution and biology on the planet. Full stop.

Secondly, it's also the start of capitalism and the start of this sort of "Wild West frontier" of raiding different continents, and so forth, literally leads through to the present day.

And the third thing is, we believe that the annexation of the Americas and depopulation—not deliberate, but this unfortunate depopulation—allowed Europe to industrialize.

So, it's really interesting if you think about it: China has an empire for 5,000 years. It doesn't have an industrial revolution. The Ottoman Empire existed for hundreds of years: no industrial revolution. Egypt has a 4,000–5,000 year empire, also no industrial revolution. But Europe does.

And if you've visited Europe in the medieval times and looked at it—well, I'm sorry but if you compare it to the Chinese Empire or the Ottoman Empire, it looks like a cesspit. I mean, really, it's a backwater. There's no way you would turn around and go, "***This** is going to be where everything changes to drive the whole world economy.*"

But the problem with all the other empires is that you have to have a certain number of people working on the fields to produce enough food for everybody. So, you're all landlocked: you're locked to a certain amount of food and a certain number of people.

And if you have a whole new continent stripped bare, from which you can steal the gold, the silver, all their resources, all their food and ship it to your continent, you suddenly can break that population boundary.

And that happened in Europe. At the same time, you have European countries all about the same size, all of which had kings and queens who were related and in deep competition with each other.

Over the last 800 years, the major European powers have spent about 50% of the time at war with each other, and war drives innovation. So, you suddenly go from the Chinese having gunpowder to the Europeans producing some of the most accurate rifles on the planet and being able to fire a cannonball over a mile due to this enormous competition.

Then Britain invents the steam engine to pump water out of the mines in Cornwall so they can dig out more tin for their industry. And then, of course, we suddenly run out of wood because we deforested from building ships to keep Napoleon and the French out. And so again, that environmental limit meant that we switched to coal. And then, presto, you have the Industrial Revolution.

HB: And you had coal, as it happens.

MM: Yes. So, all these things had to be in place. Without the annexation of the Americas, you would never have had an Industrial Revolution in Europe that started in Britain because of colonialism and fighting between other European powers. And therefore you would never have had The Great Acceleration and you would never have had where we've got to now. So, that's why we put the starting point to the Anthropocene at 1610.

Questions for Discussion:

*1. Do you agree or disagree with Mark's historical analysis of what caused the Industrial Revolution? Readers interested in a different perspective on this topic are referred to **Enlightened Entrepreneurialism** with UCLA historian Margaret Jacob.*

2. Do you think the term "Anthropocene" is a helpful or unhelpful concept?

III. What We Know

Ice ages, snowballs and hockey sticks

HB: I'd like to go back to a much earlier era in terms of climate. Because when people start thinking about climate change, an obvious thought is, *Well, hang on. Weren't there all these ice ages that have happened in the past? Hasn't our climate been changing regularly over millions of years?*

My rough sense of things is that ice ages happen over something like a 100,000 year cycle, which is caused by small changes in the Earth's orbit around the sun in terms of its eccentricity and precession and so forth. Is that a fair summary? And when did we get a clear understanding of all of that?

MM: The seminal paper on ice ages came out in 1976—it was by Jim Hays, John Imbrie, and Nick Shackleton, my PhD supervisor ("Variations in the Earth's Orbit: Pacemaker of the Ice Ages"). As it happens, I've written an "In Retrospect" piece for *Nature* on that article ("Forty years of linking orbits to ice ages", December 7, 2016).

The first mathematical framework of the concept was developed by Milutin Milankovic in 1941, but nobody had the ability to test it. So what they did was they took long, deep sea sediment records. They then measured different proxy records in the records: one of them was the fossils assemblage that allows you to calculate the temperature of the water. There was also oxygen isotopes allowing you to give an estimate of how much ice there was on the planet. And that gives you nice cycles and shows you the big ice ages.

What they're able to do is statistically link those to the beautiful wobbles that you described, which are precession, eccentricity and obliquity. Those wobbles of the earth change the amount of the sun's energy hitting different parts of the planet. And the key one is actually Northern Hemisphere summer. If those wobbles make summers just a little bit cooler, a little bit of ice can survive. And if that ice gets through one summer it will survive through the winter and grow and get bigger and bigger. And that's how you build up an ice sheet: you can start to build that base just by having that summer slightly cooler, because you get less solar energy.

So, for about the last 40 years, we've known the underlying causes. What has happened in the interim is that we've understood, firstly, the resolution: we've actually now got such high resolution records, we can see how quickly the ice ages take. And actually ice ages take a *huge* amount of time to build up. So, over that 100,000 year cycle, you're looking at perhaps 80,000 years to get the full ice age.

One of the key things we've done over the last 40 years is improve the resolution of our records to see how quickly ice ages form and how quickly they disappear. And we've seen that in the last eight cycles or so, what happens is that you have these slow buildups of ice, which takes about 80,000 years to get a full ice age—that's when you have ice sheets three kilometers thick over North America. We're talking about a *huge* amount of ice here.

And these ice sheets expand from both poles. In North America, they come down to The Great Lakes. And in Northern Europe, you find that the ice sheets actually stopped just north of where we are now in Finchley, where my mother-in-law actually lives. It's called the Finchley depression—not because of my mother-in-law—but it is a depression where the ice sheet before last actually stopped. And so you have most of Northern Britain covered in ice sheet and Fenno-Scandinavia as well.

So you have these big ice sheets. To give you an idea how much ice was locked up, the ocean—which covers 70% of the world's surface—dropped by 120 meters. So, that gives you an idea of the amount of water that was locked up in this ice.

But we also know that at the end of an ice age, it takes less than 4,000 years to get rid of all of that ice because they have very positive feedbacks. Cooling down the planet takes a long time because ice sheets aren't stable: they like to collapse. But warming up the planet is actually quite easy since it evolves very positive feedbacks.

We also know by looking back, we can see where we are now in the actual story. So, we're currently in the Holocene, which is an interglacial period.

We can look back at previous interglacial periods and see that for the last two and a half million years, our warm period of time makes up about 10% of that period. And so we're in the warm, slightly unusual period when we don't have ice sheets.

I think it was James Lovelock who wonderfully said, "*Through climate change, all we're doing is adding to the fevered state of the planet.*"

So, we're just adding to the warmth that is a bit unusual for our period of time. We're actually in what is called an "icehouse world": we're quite ice-locked at the moment. We're in the only geological period of time that I know whereby we have ice at both poles.

Usually, in big ice ages and the geological past, we have big ice at one pole but we haven't had ice at the other pole. This is one of the few times we have ice at both poles. If you go back to when the dinosaurs were roaming around the planet, it was much warmer, much lusher. We had rainforests up to the Canadian border down to Patagonia. However, the difference there is that there weren't huge amounts of ice lying around waiting to melt and actually cause problems for us.

HB: Before we move to some of those problems, I thought I'd make a bit of a diversion and talk about "Snowball Earth" that some people might have heard about. What's that about?

MM: When people talk about Snowball Earth, we are talking now one to two billion years ago—way back in deep, deep time, almost before real geology starts. Geology usually starts with the Cambrian explosion of life, which is about 550 million years ago, but Snowball Earth is before that.

What happens is that because of feedbacks that start to cool the planet down, the ice starts to grow and there isn't anything limiting the actual ice growth. Nowadays, we have lots of plankton in the ocean that build carbonate shells, so we have a buffering system. Which means that even though CO_2 can go up and go down, it's geologically bound by particular conditions because of the buffering of the ocean. But those didn't exist back then, so there was nothing to stop the whole system cooling down.

And there's a huge discussion about whether we had a Snowball Earth—the whole planet was ice locked—or a sort of Slushball, when the oceans were still able to move. The problem that we have with a Snowball Earth is that we can see how you can get into that catastrophe, but it's less clear how you can get out of it. How do you get the system to warm up? Some people suggest that perhaps with volcanic eruptions, enough CO_2 and methane are released into the atmosphere to then start the warming process that would undo things.

HB: And as you say, that's highly non-linear. So once it starts it...

MM: You can then get through it.

But what's interesting is that as soon as you get protozoa evolving in the ocean that make calcium carbonate shells, that starts to buffer the whole carbon system. We've never had a Snowball Earth since. So even the great ice ages that we've had during the Carboniferous period and 18,000 years ago, all only get to a certain level. They never get any closer to the tropics than, say, North London.

HB: Interesting. So let's return to where we are now. You mentioned that we are currently in an interglacial period. To what extent, and with what level of accuracy, have people been able to calculate just how far we have postponed the next ice age, the oncoming ice age because of the extra CO_2 that's now been inserted into the atmosphere?

MM: There have been some very good studies on the future of glacial-interglacial cycles, because the other thing we've learned in the last 40 years is that the wobbles just push the climate in a particular direction.

So when my students say, *"Oh, yes, orbital forcing causes ice age,"* they're wrong because that isn't the cause: it's the climate **feedbacks** that actually cause it.

For example, we have the same configuration today as we had 18,000 years ago, but we're not in an ice age. So where you've come from depends on which way the orbital forcing pushes you.

By looking at other interglacials, what you find is that in all the other interglacials what you get is deglaciation: all the ice disappears, CO_2 goes up, methane goes up. They stay high and then they just slowly drop down to about a certain critical limit—which is usually about 240 parts per million for CO_2—and then we slide into the next ice age.

OK, that still takes a long time, but we turn that corner. What colleagues have seen is that this very pattern—with the CO_2 and methane gradually levelling off—was happening at the start of the Holocene, but then about 5,000 years ago, methane and CO_2 start to increase—which (as I mentioned earlier) we think is because of early farmers, the deforestation, the cumulative human impact.

And so instead of traipsing down to that 240 critical limit, it's risen instead to 280 by the pre-industrial age. So you can see that, perhaps around today or within a thousand years or so of today, we should have started to go into the next ice age.

But because of that rise, we've delayed the next ice age; and with the amount of carbon we're putting into the atmosphere now some estimates suggest that we've definitely put off the next ice age for another 50,000 years.

And if we carry on as business as usual with the carbon, then we'll probably put it off for about another half a million years. So human impact is so large that we're now significantly minimizing the effects of celestial mechanics, basically saying, *"No, the climate system is not going to react."* We could be delaying the next ice age for half a million years.

HB: Tell me a bit more about the actual measurements that we're making to base such conclusions on, particularly in terms of what has changed in the last 100-150 years or so. How do they work and why should we have such confidence in them?

MM: In terms of the last 150 years or so, we're very lucky because we have great records. The Victorians did a wonderful job measuring the world around them. The great thing about temperature is that thermometers are incredibly accurate and incredibly easy to use, so we have huge numbers of records from around the world.

We also have a lot of sea temperatures as well, because all commercial shipping is mandated to take measurements of sea temperature and other things as they are moving about across the oceans. So we have this huge amount of data.

And what different organizations have done—and this is important, that different organizations have done it: we have NASA, NOAH, the Met Office here. We even had a rogue group of physicists in Berkeley who didn't believe the temperature records so they did it themselves.

HB: That was this guy Muller, right?

MM: That's right: Richard Muller. He didn't believe the average temperature records, so he decided to perform the calculations independently.

And when you put all these together—you have to be very careful because different places measure things differently, so you have to correct for that—what you see is that for the last 150 years there is a strong warming. You have a slight hiatus during the '40s. Then you have a strong warming again through to the present day.

Now that's built on all these climate records put together: the Japanese Meteorological Agency, the Met Office, NASA, NOAH. All those records look very similar. They're different in the detail, which is great because they're doing different stats, they're compiling it in different ways.

And what was really interesting is that when Muller and his physicists put theirs together, they actually got a *greater* temperature range than others did. Many suspect that this was because they needed to further correct some of their records, but the net result is that many different, independent groups have confirmed that we've warmed up the planet a bit less than 1° Celsius over the past 150 years (0.8°C–0.9°C).

So, that's for the past 150 years, but we can also look longer term. We can look at ice core records from Greenland where the ice accumulates very quickly and into glaciers. And you can look at proxies for temperature and you can then splice those together.

This is where we get what is commonly called "the hockey stick", because you can reconstruct temperatures going back some 4,000–6,000 years. You can see that the temperatures have been slowly cooling down. And then we have this big tail or big stick at the end, which is where *we've* actually then started to influence greenhouse gas composition and therefore increased temperatures markedly.

Questions for Discussion:

1. *Why do you think that creatures in the ocean that produce calcium carbonate can act as a "buffer" against a Snowball Earth scenario? Readers interested in more detail on calcium carbonate producing creatures are referred to* **Coral Reefs: Science and Survival** *with Charles Sheppard.*

2. *How does the story of Richard Muller illustrate an important difference between scientific scepticism and other types of "climate-change deniers"?*

IV. Unchecked Opinion

Examining beliefs

HB: I can imagine somebody watching this and saying, *"OK, I'm willing to accept the fact that humans are changing the climate by their actions and I'm fine with calling that 'the Anthropocene' if you'd like. But so what? I don't really care that the next ice age has been put off for a while because I'm not really keen on my remote descendants having to deal with an ice age anyway. So why should I particularly care about this?"*

MM: Well, the worry about climate change is not about the next ice age. So let's ignore that. That's an "academic discussion" because, as far as we know, it's not going to happen.

What's important about climate change is that everything we do in society is based on our knowledge of climate. When farmers plant seeds, when farmers plow, when they then harvest—all of that is based on their knowledge of rainfall, when it occurs, temperature, growing season and so forth.

If we then look at our infrastructure: the way we build our railways, our roads, all of that is based on our knowledge of what the range of temperature is going to be.

For example, if you're in London, you'd only have air conditioning if you happen to be in an office block. If you're in a normal person's home, you won't have air conditioning because we don't have temperatures that are that hot.

On the other hand, if I'm going to Houston, I'm not going to stay anywhere that doesn't have air conditioning. It's because we know what the range of temperatures are.

This is something I always say to my students: humans live in almost every condition on the planet—from Inuits who are happy living at minus 10 and get stressed when the temperature gets to zero because it's too warm for them, to people living in Sub-Saharan Africa or Texas when 40–45 degrees Celsius is just a balmy summer's day.

We know what the boundaries are. But what happens with climate change is that, because we're putting more energy into the system, not only are temperatures increasing, but the amount of extreme weather events we have is also increasing: rainfall increases and it happens in more intense bursts.

It also means there's more flooding, therefore more droughts. So we have those extremes to deal with. And if we were all equally rich—if we were all like wealthy Europeans or North Americans—we might not be as concerned about it apart from the big hurricanes coming in and big disasters like that, because we could adapt to it.

But the majority of people in the world are subsistence farmers. They're producing food in the tropics, and what they don't use themselves they sell to feed the people in the cities. In fact, most of the population of this planet is very sensitive to changes in climate and weather. And that's the insidious nature of climate change, it is just changing the rules whereby we live—and all of our calculations are based on normal weather and that's not going to happen in the future.

HB: OK, so let's talk about some of the politics of this. One of the interesting things about climate change is that it touches on so many different areas: science, politics, economics, anthropology and many more. So I have a theory that I'd like to bounce off you to see if you agree or disagree.

MM: OK.

HB: My focus is on the way things are portrayed. What I mean by that is that I think that many people—perhaps even most people—have a hard time understanding the process of science.

Of course there are those who have their vested interests—people who work in the fossil fuel industry, say—but let's forget about them for the time being. I'd like to talk about your average guy on the street, or even your average honest journalist, who is trying to get a handle on what's going on.

They're getting all this verbiage coming at them from different directions, and they're hearing all these angry voices, and what I think is happening is that most people interpret all information they receive within the context of there necessarily being two sides to every story.

This is, of course, a well-established and well-understood context for political issues. Every time I hear a politician saying, "*X is the way to go,*" I know I can always find her political opponent saying something like, "*That's nonsense: X would be disastrous.*"

So there's this idiom that is universally used to processing information. Which means that when a group of scientists stand up and say that there's some scientific consensus that has been reached on some issue, the media will immediately say to themselves, "*Right, that's one side of the story. Let's now look for someone else who represents the other side and give them equal time in the interest of fairness and journalistic responsibility.*"

But while that is probably a completely reasonable approach in the political realm, it is a disastrous approach in the scientific one. And most people don't appreciate that there is an important difference between the two realms. So that's my theory. What do you think?

MM: I think that you've picked up on a really interesting point, which is how the media portray science. And the media does, unfortunately, fall back into the "for and against/pro and con/have two equal people", which has been something we have been fighting against for the last 25 years—not just climate scientists, but many different scientific areas. And I think that's important.

However, I don't think that's why people find climate change difficult to accept. Most people have at least some high school science, and the way science is taught in schools is not contentious. And if you look at trust ratings in the Western world, scientists still come out as one of the most trusted professions, above doctors in fact.

So the whole idea that science is contentious is not supported by the evidence. But what people do have is their own political views. And what a lot of sociological studies have now shown is that the person's personal politics affects what they can or cannot believe.

So if you unpack climate change, you realize that it is basically a pollution issue: we are putting carbon into the atmosphere through burning fossil fuels and by deforesting areas. So we have to change that. But to do that, we have to act for the greater good. We have to work with other countries. We have to work collectively.

That actually is diametrically opposed to the major political thinking of the last 35 years, which is neoliberalism. Neoliberalism emerged out of the '70s, out of capitalism. And with the fall of the Soviet Union, the rampant idea of individual freedom, the markets know best, was cut loose and allowed to develop through Margaret Thatcher and Reagan.

Now the problem is that neoliberalism says, "*Okay, so individuals have to have the right to do what they want. Markets know best. We want to reduce government interference because that red tape slows everybody down. And this is for the greater good, because if you make money up here it all trickles down to the poor people and everybody gets lifted out of poverty.*" Even the IMF have now said, "*Perhaps the stated aims of neoliberalism haven't actually occurred.*" So there is now major economic rethinking, "*Well, hang on the trickle-down hasn't worked.*"

But my point is that if you happen to believe that, *Markets are the best control, governments are necessarily interfering and problematic*, the notion of environmental regulations to control the amount of pollution you can put into the atmosphere that has to be agreed upon not only within your sovereign country but also internationally—well, that's something you're naturally going to oppose.

And a lot of studies have shown that it doesn't matter how much scientific evidence you present to certain people. It doesn't matter if you back it up with more and more studies. Often scientists get into the whole trap of, "*Perhaps I haven't got enough evidence—I need to get more,*" and then piling up more and more evidence.

But there are certain people who, just because of their fundamental political principles, cannot accept climate change. They have to deny it, otherwise it undermines their political views.

HB: Here's my problem with what I believe you just said. It seems to me that what you're saying is, "*Yes, people understand science because they have some high school scientific education and they understand basic science principles.*" The claim seems to be that there's some understanding of science and then, parallel to that, there's a sense of—

MM: Of trust.

HB: Right: trust. And then you go on to say that many people effectively think to themselves something like, *But if this clashes with my political principles or my economic values or my right to go out and make as much money as I possibly can, then I'm simply not going to believe it.*

So I would say that adopting such a belief clearly demonstrates that you ***don't*** understand science at all, because science doesn't care whether you agree with it, whether you like it, whether you dislike it.

Put another way—I'm not sure I can come up with a particularly worthy analogy on the spur of the moment—but it would be akin to saying something like, "Well, the way those physicists tell me that apples fall interferes with how I feel about free will and macro-economic forces. So I disagree with them. Sure, I know they have their theories and their data and their centuries-old tradition, but I don't care because their way of telling me how apples fall inhibits me from feeling sufficiently confident to start my multinational corporation. So I'm just going to say, "*They're wrong—it doesn't happen that way.*"

And, to me, that shows that you don't understand the process of science.

MM: The interesting thing is then what you would expect is as someone's scientific knowledge grows, you'd expect that their acceptance of climate change would increase.

However, the beautiful studies coming out of Yale showed that depending on which party you were most likely to vote for in the United States was the key factor and not your background in science.

If you happen to be more oriented towards the Republican Party and you had a high level of science, you were nonetheless still less likely to believe climate change, because you would say, "*Well, I know science, I've done science to a degree level. Therefore, I know these guys are actually making it up.*"

So, again, the problem is that science is being used against science to actually allow people to continue to believe their own world view, which is really important for individuals.

HB: So I'm not sure we're having an argument, but perhaps we are.

I would say again that my conclusion is that there is a real education problem. I'm not saying that those guys at Yale got their study wrong—I'm willing to bet that their results are completely accurate. But I would simply say that I don't care if you have an undergraduate degree or not, anyone who views a scientific issue within a political paradigm simply doesn't understand science. Full stop.

As it happens, I've met many scientists—I daresay to venture, far more than you even have; I'm older than you are and ran a research institute for several years—who are obdurate, emotional, irrational and even flat-out wrong.

But the idea that an *entire community* of global scientists would repeatedly and consistently fabricate their conclusions in such an overwhelmingly data-driven field is laughably wrong.

And if you think that such a thing is actually the case, you simply don't know what science is and how it works. Anyone who believes in some sort of global conspiracy from a body like the intergovernmental panel on climate change is completely out to lunch. Period. And if it turns out that she has an undergraduate degree in science on top of that, then she clearly hasn't gotten very good value for her educational dollar.

MM: But we're in a new political age. We're in the post-truth age. People want to believe what they hear. I mean, we've got politicians going, "*We do not need experts*" on both sides of the Atlantic.

There's this whole attitude of, "*Experts? Well, they don't know what they're talking about. We don't need experts. Why should somebody train for 20–30 years to be able to tell me something important?*" Meanwhile, there's a sense of, "*OK, I might accept my doctor's prognosis because I really care about me and I might take that, but anything else I'm not going to accept.*"

So we've got to this really weird, strange period of time in the early 21st century whereby the populace, and particularly politicians, are now stepping back. They're not saying, "*Well, my experts say this. Your experts say that, but I don't believe them.*" They're not using experts against each other. Instead they're saying, "*We don't need experts. I'm going to tell you what you want to hear.*"

So it's not just science that's suddenly having this happen. This is starting to actually ripple through many different sectors of our society. Some of my colleagues would trace it back to the idea that science is "a belief system".

I ask my students about this at the very outset: there are 150 first year students and I give them a lecture on climate change.

And the first thing I say is, "*Put your hands up if you believe in climate change.*" So, of course, almost all of them put their hands up. There's always one contrary because you have to have some independent-thinking student in the middle muttering angrily.

But then I tell them, "*You're all wrong.*"

Because science, as you quite rightly said, is **not** a belief system. It's a rational system of collecting data, collecting evidence, making theories, testing those theories—both in the real world and in labs—and then collecting more data to perform a rational analysis of what the world is and how it is going to change.

HB: Exactly.

MM: Again, it's not a belief system. So when my students say, "*Yeah, I believe in climate change.*" I tell them, "*No, I'm sorry. It's not a belief* *system. You can trust. It's a trust system. Do you trust the experts? Do you trust the science?"*

HB: Well, that's a shame—I was looking forward to a bit of an argument, but I don't think we can do that because I essentially agree with you. I must say, though, that I'm quite flummoxed as to how we got here. Much as I'd like to, blaming neoliberals for everything seems a bit over the top to me. You, on the other hand, don't seem to have any qualms about blaming neoliberals for everything.

MM: No.

HB: So let me say something else instead. Let's single out the United States and United Kingdom, two wealthy, industrialized countries. In fact, let me focus on the United States, which is a very strange place, because mixed in with all of this anti-science, anti-rational thinking, categorizing objective evidence as "belief systems" and so forth, it's also the greatest place on the planet for scientific innovation—

MM: Oh, absolutely.

HB: —and high-level education and scholarship in such a wide variety of different fields. So how did such a bizarre juxtaposition happen? How did we get here? I mean, it's a bit of a stretch to blame everything on Milton Friedman in my view.

MM: Again, what is, I think, both exciting and scary about the 21st century, is the ability for everyone—and I mean *everyone*—to access information. You can be in the middle of Kenya chatting to some of the locals ,and they say, "*Oh, hang on,*" get their smartphones out and check. They didn't even bother with the computer age—they missed that completely and went straight to the smartphones. It's fantastic.

But the problem there is if every individual has access to gazillions of data bytes of information and opinion, how do you sort out what to believe? Which sources do you trust? Now, of course, there are some who have, basically, gone through the education system and have a critical ability to think, to work out who has an agenda for what. But most people don't have time to do that.

When I take my kids to school and look around at the other parents, most of them are lovely but they're just about keeping their heads above water. They dump their kids at school, they go to do a hard job all day to make sure they have enough money to try to give them a better life before picking them up again.

They don't have time, like us academics, to sit back and say, "*I'm going to think now. I'm going to think and pontificate about who's actually telling me porky pies and who's telling me the truth.*" They don't have time for that. So they get bombarded with all this information and, therefore, they use their gut or their friendship circle. What's interesting is people's beliefs are usually controlled by who they know and who they actually talk to.

For me, it's all about the ability to process that huge amount of information. 50,000 years ago, we were able to actually work out what our tribe was doing and who was doing what to whom, and so forth.

But now we have the ability to check on the opinions of, basically, every one of seven billion persons on the planet if you really want to. How do you process that?

I think that's what the 21st century is about, learning how to deal with that. And we don't have the political systems that allow that to actually work.

Questions for Discussion

1. Do you think more time should be spent teaching the process of science rather than simply the "facts" of science? If so, how should that be done, exactly? If not, why not?

2. Do you think that there's a clear difference between "a trust system" and "a belief system"? How would you distinguish between them?

3. Do you think that Americans are more likely than others to view issues like climate change within a political context? If so, why do you think that is? For additional perspectives on this issue, see **Saving the World at Business School Part 1** *and* **Part 2** *with University of Michigan business professor Andy Hoffman, as well as Howard's 2021 book,* **Exceptionally Upsetting***.*

V. Planetary Perspectives

Which type of Anthropocene do we want?

HB: So let me ask a potentially more provocative question. So if you were not an academic who had the time to think, as you say, but instead if you found yourself prime minister—and let's say prime minister with a majority government, so you could do effectively, whatever you wanted.

What would you do regarding this issue? And by that I mean not only making people understand the ramifications of climate change and the cultural change that needs to be accompanied with that, but also those broader-based issues we were just discussing: educating them about the process of science and what scientific thinking means.

MM: So the first thing is that, as I teach my students, if you happen to be a politician, you have multiple stresses on you at any one time. It's very easy as a climatologist to say, "*If you're going to deal with climate change you need do this and that.*"

But as I say at the beginning of all my climate change lectures, "There are four big global challenges for the 21st Century: global poverty or inequality, depending on how you wish to define it; climate change; environmental degradation, which is happening even independent of climate change, and then there's global security.

So whether you're in control of a country or a region, you have to be able to provide solutions for all four and make sure that your solution to one doesn't actually exacerbate any of the others. So it's not an easy job being a politician, although most academics think, "*Oh, it's easy. Just do this.*"

HB: Well, most academics think any job other than their own is easy. Which is a statement which probably generalizes to other professions as well.

MM: That's right. But having said that, if I was in charge of the British Isles, the first thing I'd look to do is focus on shifting the subsidies. So of course I'd shuffle the subsidies towards green energy and renewables—firstly because, as a British subject, I do not want to be burdened with the idea that Russian gas, Kuwaiti gas and oil from the Middle East is going to dominate and control my economy.

I'd also want to actually make sure that my North Sea oil, which is a wonderful revenue generator for the British Isles, is used up as slowly as possible to maintain it for the future. So, keeping the fossil fuels going from the North Sea to make sure there's revenue, but at the same time shifting us to a low-carbon future.

Now, the other thing I'd be doing is ensuring that we were the leaders in that technology because we *were* the leaders in wind turbines until the Danish bought the companies and moved them to Denmark. Now it's Denmark—and, interestingly enough, China—who are the leaders in developing wind turbines.

I know, and can see from all of our papers, the growth in the green economy globally. And I want to be at the forefront of that. I don't want to be saying, *"What do you mean I can't have a green car that goes 200 miles an hour?"* I want to make sure that this incredibly technologically advanced country is right at the leading edge.

So I think those things are very, very important. And then you try to leverage that success globally to positively influence other countries and other regions, because there is nothing difficult about what we're trying to do.

People think it's expensive. It isn't. The reason why fossil fuels at this moment in time seem to be cheaper than renewables—

HB: —is because of subsidies.

MM: That's right. The IMF, the International Monetary Fund, has calculated that with the climate-change damage put in, global subsidies to the fossil-fuel industry amount to about $5 trillion per year. That's twice the GDP of the United Kingdom, which is the fifth largest economy in the world. So it's a huge amount.

Now, the reason being is not due to neoliberalism; this is not actually due to all those private companies making huge amounts of money. It is, in fact, about state-owned companies.

If you look at the top 25 fossil fuel companies in the world, 19 of them are fully-owned or partly-owned by a country. They get huge tax subsidies, they get huge tax breaks and direct subsidies and all of that to allow them to produce fossil fuels as cheaply as possible.

So what we need to do is actually make sure those countries are forced to reduce those subsidies to actually make renewable energy much more effective.

And so, while I do complain about neoliberalism, in this case it's actually state-owned companies which are getting unfair advantages, which are skewing the whole market, which means that our green and clean tech isn't getting the opportunities they need.

HB: You've clearly thought a lot about this. Are you contemplating running for office at some point in the future?

MM: Many years ago I'd done a lot of work with arts groups and the creative industries in looking at how to discuss climate change and sustainability issues. Generally speaking scientists have this certain way of talking that doesn't actually inspire you, whereas a painting or a statue or a film can just blow people's minds.

But at one of these events, during the Q&A, one of the people asked, *"What would you do if you were in charge of a small African country?"*

And my instant response was, *"Right. So I've got a poor country. Now, I'm pretty sure I've got some coal reserves in my country; I'm just going to assume that. So I'm going to build coal-fired power stations as quickly as possible to produce cheap electricity for my country to lift my country out of poverty."*

And they looked at me, shocked, and asked, *"But what about climate change?"*

"Well, you didn't ask me about that. You asked me what I'd do for my people, for my small African country because I'm now the dictator. I'm making sure that they're fine and safe."

At which point they went, *"Ah."*

And I said, *"Yes. That is the problem."*

Because if you ask about a particular solution, that may not be the same as the one you're focused on for climate change. And that does worry me—a lot of climate change scientists only see climate change as the issue. Whereas as a geographer, I am aware that there are so many other issues that we have to deal with.

And the thing is, I'm optimistic because in the first half of the 20th century, we had to deal with huge problems of lack of representation, small countries being periodically just swallowed up by bigger ones, the end of colonialism and massive wars.

At the Bretton Woods conference in 1944, they sat down, brought out their desk-drawer plans and said, *"This is the new world order: the United Nations, the World Bank, the International Monetary Fund,"* which then produced an *amazing* increase in people's longevity—we doubled the population. All the things that were problematic in the first half of the 20th century were fixed in the second half, amazingly.

However, that's led to another set of problems: poverty, inequality, environmental degradation, climate change and security issues, which we now have to actually try to unpick and deal with in the first half of the 21st century.

HB: So let's talk about tackling those issues. We spoke earlier about people who I would describe as not having a sufficient understanding of the rational principles and processes of science. You spoke about the dangers of looking at science as a belief system, which leads to a dangerous correspondence with politics and scientific conclusions. So there's that.

But then there are those who have a different view, one I would describe as having "too much faith" in science and technology, saying things like, *"I'm not worried because whatever size mess we make, I'm certain that those clever scientists will be able to fix it. They'll find a way to geo-engineer the planet and return everything back to normal so we can carry on as business as usual and when things get really bad the scientists will intervene."* How would you respond to those people?

MM: For me, this is why even the discussion of the Anthropocene is really important.

Climate scientists have become louder and much more insistent over the last 25 years about climate change because they felt that nobody's listening. So they keep pushing and pushing and pushing.

Unfortunately, what that's done is to allow other things that are happening due to human activity to be missed: the huge loss of biodiversity, the overwhelming amount of deforestation, the amount of changes in natural habitats, our alteration of even the nitrogen cycle—because, at the moment, half the nitrogen fixed from the atmosphere is fixed by us to make fertilizers, to produce food—*all* of these things, which show that our impact on the planet is much wider than only changing the climate, have been missed.

So this is why I'm really excited about the discussion about the Anthropocene, which effectively says to people, *Look, we are a geological superpower.*

Now I appreciate that for the last 500 years or so scientists seemed determined to make us feel increasingly insignificant. From Copernicus informing us that the sun was at the centre of things rather than the Earth to modern cosmologists informing us of how that sun is actually a very insignificant little ordinary star in an enormous universe.

And then of course the biologists were not to be outdone, with Darwin reducing us from our privileged position between animals and angels, informing us that, actually, we're just an ape that lost its tail and a bit smarter.

So for the last 500 years, the fairly constant scientific message has been that individual humans are tiny and insignificant: it's a big, dangerous, nasty universe out there, and we're just one small part of nature that has evolved.

However, the Anthropocene says, *Well, hang on; collectively as nearly seven and a half billion, we're actually the most important thing on this planet, which is the only planet where we know life exists in the whole universe; and we control its destiny.*

And this prompts us to ask, *OK, then: what sort of future do we want for this planet?* People are now starting to talk about "bad Anthropocene" and "good Anthropocene"—contemplating different possible futures that we collectively can imagine for the future: whether as custodians of the planet we make sure that we keep everything as nice as possible for all of us—which means that we actually look after all of us and the planet—or do we just have business as usual and keep going as we are because, well, that's just what we do.

For me, then, the big thing is that in the early 21st century we've now hit that choice of which direction to go. The big problem is: How *do* you organize seven and a half billion people who are used to working in small groups, because of our brain, of about 150? How do you make that step up from a small tribal creature, which we are, to a mega-creature that covers the whole planet? I think that's the challenge that we're going to have this century.

HB: To that end, it seems to me that there are some genuine reasons for optimism, namely the Paris Accords that were recently signed. Does that give you some sense of optimism? Do you think, *Maybe there really is a possibility for a good Anthropocene; maybe we can actually get there as a global society?*

MM: I think Paris was absolutely essential. It was key that it had to work, particularly in light of the fact that the 2009 meeting in Copenhagen completely failed. And the levels of expectation and optimism for Copenhagen was so high beforehand that the outcome just shattered the whole of the negotiations; and a great deal of time and effort had to be expended to build it up again.

Now the key thing about Paris, I have to say, is that you have to give credit to the French, which is a difficult thing to be doing.

HB: Not for me. I live there.

MM: Well, it's very difficult for a Brit. But what they were able to do was diplomatically make countries discuss things and actually come to an agreement. Or even if they didn't come to an agreement, they got told that they had come to an agreement, and then realized that they couldn't turn around and say, "*No, we didn't agree to this.*"

So some very interesting politics were being played there. Some colleagues of mine were in the heart of this and saw it going on. It wasn't the French themselves who would do it, but they would get someone else—a relevant president or prime minister or key influencer to phone up and start the ball moving.

It's a watershed, because what it says is all the leaders of the world—including Saudi Arabia, including the United States—see climate change as a problem and that something must be done. Now, the pledges that have been made for Paris, aren't nearly enough to get us to two degrees, but it's the first time we have real, new pledges.

HB: And a genuine global acknowledgment of the problem.

MM: That's right. And also we have a name, which is the two degree target. Now that's completely politically made up: an aim doesn't necessarily have any logic within science, because if you happen to live in a small island state, say Tuvalu in the middle of the Pacific, guess what: two degrees means your whole country is going to be flooded. So it's artificial, but it gives politicians and people something to look at and aim for, to know whether we're actually doing something good or bad.

I think what we do need to do however, is underneath that, we need to change economics. And I don't necessarily mean economics as in the day-to-day exchange of money, but how we actually teach economics, and the big problem is that economics is the gold standard.

At the actual negotiations, we know that the people we have to get to aren't the actual environment ministers, it's the finance ministers, because they're the ones looking at the bottom dollar and basically working out whether they can actually afford Paris or not, because they are looking at different stats.

So one of the interesting things that people have done, and this is where we were involved at UCL, is associating health with climate change, because one of the big costs for every single country is health-care. In the UK, we spend about 10% of GDP on healthcare. In the USA it's 17% of their GDP.

It's a huge thing. So you go to the health people and say, "*Look, this is what's going to happen with climate change. This is going to increase your health care spending. It's going to cause this problem and that problem; these diseases are going to spread out.*"

They then go to the finance minister and say, "*Do you know how much extra this is going to cost us in 10, 20, 30 years?*" That then gets the finance person worried, who then starts to pay serious attention.

At the moment, what we're hearing is the change in economics: the idea that the old classical economic models don't actually work and we do need new economics. However, getting the economics departments of every university in the world to actually move into the 21st century is difficult and problematic. For some reason, unlike science departments, they don't seem to learn.

HB: They're not bad at retrodiction, though, you have to give them that: they're reasonably good at predicting the past.

MM: Yes. And again, I have no problem with them getting the future wrong because it's very difficult to predict people in the future, but their underlying principles—i.e. everybody's rational, everybody makes these decisions based on cost benefit analysis—just don't work in the real world.

And I think we need realistic economics for the 21st century so we can make realistic choices. And the other thing that always frustrates me is that all the natural environment is never costed in. It's a freebie. It's a freebie, even though now they are saying, "*Well, perhaps we actually have to bring in climate change damage somehow.*"

This is where Nick Stern's report ("The Economics of Climate Change: The Stern Review," published in 2006) was absolutely amazing, because what he said was, "*Look, you do it now and it will cost 1–2% of world GDP to fix climate change. You leave it to the middle of the century, it's going to cost you 20% of world GDP, possibly more. Which one do you want?"*

Now the economists complained that he didn't do a discount because he basically said that a pound or dollar now is worth the same in the future. But he effectively said, *"Fine: morally, the future and today cost the same. Now, what are you going to do?"*

And that was instrumental. That did change a lot of people's thinking about climate change, with some starting to think, *Well, perhaps we should change things now.*

And the other thing I always get frustrated about is that whenever I meet the top business leaders in any country, I'm constantly struck by the fact that the one thing they want is consistent regulation. They tell me, *"Look Mark, I'm really good at what I do. I can make money all the time. As long as the rules stay the same."*

What they also dislike is if Joe Bloggs down the road is allowed to get away with it and making more money than them because he's not following the rules but not being caught out. They get really frustrated by that. They want a level playing field.

They don't mind environmental regulation. That's fine. As long as they are consistent and applied across the board so people are caught out who are trying to cheat it. The best of the best business will always make huge amounts of money because that's what they're good at. But politicians don't see that, they don't understand that.

Questions for Discussion:

1. Should political science and environmental science now be regarded as two aspects of a common field?

2. Might insights from neuroscience help us make the transition from being a "small tribal creature" to "a mega-creature that covers the whole planet"?

3. To what extent do you think the United States significantly damaged its position of global leadership by withdrawing from the Paris Accords?

VI. Becoming Social

Investigating a watershed moment

HB: You've been very generous with your time, thank you very much. Before we conclude, I have one more question, which is a little different than the tenor of this conversation. Normally, when I talk to scientists, I ask what specific questions they would most like answered, if I were God, if I were an omniscient being.

In this discussion, naturally enough, we've talked a lot about policy. We've talked about economics, politics and sociology. You gave me a summary of our understanding of climate. We didn't talk about your work on evolution or the stability of continental slopes. We didn't talk about any of that stuff.

MM: The things I fiddle with, yes.

HB: And I'm sure a lot more besides. So it's quite conceivable that the answer to that question will have nothing to do with the prior discussion we've just had—but let's find out. If I were God and I could answer any questions, scientific or otherwise, that you might have, what might you ask me?

MM: Well, my fascination is at the moment with human evolution. I've recently written a book called the *The Cradle of Humanity: How the changing landscape of Africa made us so smart,* which puts together all of my knowledge about human evolution from the bones to the climate to the tectonics, putting it all together in a coherent story that ends up with the Anthropocene.

But the one bit in there, which I discuss a lot, for which we have no real answers, concerns the sudden cultural change that happens in our species.

Homo sapiens emerged from Africa, probably from Homo erectus, about 200,000 years ago. They walked out of Africa and they slowly made it to China. They made it into Europe. They made it even down to Australia, but didn't do anything special—just populated.

So there's nothing special there about our species for about 100,000 years. And then, about 100,000 years ago, there are a few beads that turn up, a few sort of shell engravings and so forth—OK, there's a bit of art, but not much.

But then 50,000 years after that—50,000 years ago—something changes: we get these beautiful artworks in France, in Switzerland, in Indonesia. We get these beautiful figurines you can find in Germany and in former Yugoslavia, beautiful pieces. Suddenly, culture seems to be born and takes off. And for me, the question I'd ask you if you were the omnipotent one is, *What changed there?*

At the moment, we think that we probably self-domesticated. There seem to be suggestions that, for the males: eyebrow ridges retreated, faces became shorter, finger lengths, the digits, got more equal—all of which suggests that there's now a lot less testosterone in their bodies, which means they're less reactively violent.

That doesn't stop humans being violent, because as a group humans can be incredibly violent—which is important—but individually. So under those new circumstances, if you and I bump into each other we would probably say, "*Oh, sorry*," instead of trying to kill each other.

All that changes about 50,000 years ago, we think, and that seems to be a form of domestication, which then allows us to build culture, to then learn and have a cumulative culture because there's no good someone inventing a new way of painting, or a new weapon—

HB: If it can't be capitalized on by the next generation.

MM: Yes. Being able to teach and pass things on from generation to generation is what defines us as a species. And you find that knowledge has been accelerating because we're getting better and better at passing that knowledge on to complete and utter strangers.

So for me, the question would be, *What happened? Why after 150,000 years of Homo sapiens running around, did it suddenly change?*

Because there was no climatological reason. Some people suggest there's a new wave of Homo sapiens that then comes out of Africa. But what happened in East Africa, 50,000 to 60,000 years ago that made us more of team players, more culturally attenuated?

HB: And from what you were saying, it seemed to have happened roughly simultaneously everywhere, right? Or is that not correct?

MM: Well, there are many interesting papers that have come out. Genetics is now changing all of our views of recent evolution—there's some amazing stuff coming out. What it looks like is that there's a first wave of Homo sapiens that came out of Africa about 100,000–120,000 years ago, all the way to China—there's some teeth there from about 100,000 years ago.

However, there is then another wave that comes out of Africa, somewhere between 60,000–80,000 years ago, and the genetics says that *they* replaced everybody else. So *that* is the wave that we're related to: the genetics shows that all of us are related to that wave out of Africa, not the earlier one.

HB: OK. But then, as you say, that just puts the question back to, *What is it about those people?*

MM: Yes. *What happened at that place and time to change those people, and why did that then give them a major advantage over previous humans?* So yes, that would be a question I'd very much like answered, which has nothing to do with climate change, although I'd probably argue that some of my stuff about human evolution is structurally linked to it: *the reason why we have a big brain is because we had to learn how to deal with really rapidly changing conditions in East Africa.* So even our genesis is actually due to climate change.

HB: So I suppose it's possible that the mess that we are currently making for ourselves will put more evolutionary pressure on us to become smarter, still.

MM: Well, what is really interesting is that studies of chimpanzees show that most of their behaviour is hardwired in: it's already in the genetics, in the brain.

With humans, actually, not much of our behaviour is actually hardwired in, which means as soon as you're born, the environment and society you're born into have a huge input into the way you think and how you interact. I think this is why we never understand our children, because they have completely different influences. So in a way, I reckon, every generation is a different species.

You and I are struggling with this idea of how the hell do we interact with all these people, with all these communication devices, and so forth.

But if you look at my children, they have no problem because they've been born into this. My children have no idea of a world without Google. They can't imagine not being able to google stuff. They have completely different brain patterns from us, how they think, which might yield solutions for the 21st century that we oldies can't imagine.

That, for me, is the most exciting thing about humans: because our brain is so adaptable, we can change with every single generation. We don't have to wait for hard evolutionary changes to occur. We have that brain that's so adaptable, it'll change with every generation.

HB: What you were saying reminded me of a conversation I had with a neuroscientist (*Minds and Machines* with Miguel Nicolelis) who was talking about the remarkable propensity of humans, that he believed is fundamental to our neuronal structure, to adopt new tools. This would not only explain differences between generations, it would explain differences in behaviour and attitudes and beliefs throughout a given generation.

In fact, it can work on very small timescales. If you examine where a good tennis player thinks his hand is located after playing tennis, for example, it turns out that in his judgement he's neurophysiologically incorporating the racket.

So perhaps one related question is, *What is it about human neurophysiology that enables us to actually do that?* I mean, it's one thing to point it out and say things like, My children can't imagine a world without Google. Their children, meanwhile, will immediately take Zoogle or Boogle or whatever the next thing is, completely for granted in turn. But then the question for me is, *Well that seems different than other species—so what's going on there? What is it that makes us different?*

MM: I would argue, and I argue in the book, that the thing that makes us different is that we're an ultra-social creature. So the reason we have a very large brain is not about using tools. We know that New England crows use tools, we know that chimpanzees use tools. Lots of creatures use tools.

We've recently discovered stone tools dating back 3.3 million years; so before the supposedly great increase in brand expansion, we were using stone tools. Our primitive ancestors seem to be using tools. So it's not the tools.

What's actually important is the reason we have a huge brain is because we need to track social developments. So for example, the problem is that you and I are having a conversation, and I don't know you very well and you don't know me very well.

So we basically have to have a one-to-one conversation where we're trying to think, *What's his background? Why is he asking that question? What's the reason for it?* Well, that's not too bad. You could probably get a computer to do a bit of that.

But what if your cameraman suddenly throws in a question? Then I might go, *Hang on, I don't know him—what is the relationship between these two? What's the problem?* Or you might turn around and ask, *Have you met this person? How are they related to someone else?* And that's when it starts to explode.

What I often do in my lectures—I'm really mean—is to always pick on a student, usually one who's late. So, of course, I am the alpha male and I'm therefore punishing. Okay fine, simple.

Then I turn around and ask, "*How do you like my lectures?*" At which point they hesitate. They have a problem because they don't want to appear weak in front of their peers so they might answer with a bit of humour. They'll try to get out of the situation saying something like, "*I felt much better, five minutes ago.*"

And then I'll say, "*Oh, is that your friend sitting next to you?*"

"*Oh yes.*"

"*Do you like him?*"

"*Oh yes.*"

"*OK. See the girl over there—do you think she's attractive?*"

So there's this increasing anxiety. And then I'll ask, "*Do you think she would go out with your friend?*" which further increases the tension. So again, now they're not only thinking, *What is the professor doing?* but also, *What is the girl thinking? What is the boy thinking?*

They don't know them necessarily. What is their relationship? Are they actually going out or they're not going out? Are they actually related? And then what are the other one hundred and fifty people in the audience actually thinking of them as they're trying to work that out.

And they have to do this in microseconds because they have to have a conversation with me because I'm forcing it, because I'm higher in the social standing.

That's the problem that we have as humans. So we spend a huge amount of time reassuring our social conditioning. Now, here's an interesting fact: academics spend the most amount of time gossiping than any other profession. Because it's all about status. And, of course, you can only do status by finding out what people think about you.

And I always stress that to the students. They can come in knowing quantum mechanics, or differential equations, or how to use a split infinitive. But what they don't realize is that people out there in the real world can hold the ethical or moral dilemmas of five or six soap operas in their head. They might miss it for a few weeks but then after watching a couple of episodes they know exactly who is sleeping with whom and doing what.

That's why they have such a large brain. And so I tell the students, *"Look, you may think you're slightly smarter than them, but actually that's their basis. And that's why they're so smart."*

So that's why I think that we have a very large brain, but again, how do you move that up to a much larger size than just the 150? Because 150 is about the limit. Lots of experiments have been done about how many people we actually know quite intimately, which is about 150. And the problem is when you get beyond that, you start having to drop down into stereotypes because that's the only way to deal with people, through a caricature.

HB: Well, that's very interesting, and I'm amazed it doesn't wreak havoc with your teacher evaluations—perhaps it does. Anyway, Mark, thanks very much for your time. It's been most enjoyable.

MM: Great. Pleasure.

Questions for Discussion:

1. *How might some zoologists respond to the claim, "The thing that makes us special is that we're an ultra-social creature"? For a different perspective on the social and moral tendencies of humans compared to other primates, see* **On Atheists and Bonobos** *with primatologist Frans de Waal.*

2. *To what extent is it reasonable to look for only* **one** *characteristic or behaviour or neurophysiological feature to distinguish species? Might it be the case that perceived uniqueness might result from a combination of factors?*

3. *What do you think Mark means, exactly, when he uses the word "smart"?*

4. *Might there be a way to measure "sociability" objectively? Might there be a way to somehow characterize it in terms of neuroscience? Do all humans possess the same amount of "sociability"? What determines the differences? How might studying a condition like autism help shed light on this issue?*

Continuing the Conversation

Readers interested in more details of Mark's views are referred to his many books, including: *The Human Planet: How we created the Anthropocene* (with Simon Lewis), *The Cradle of Humanity: How the changing landscape of Africa made us so smart*, and his three *Very Short Introductions: Climate Change, Climate* and *Global Warming*.

Ideas Roadshow Collections

Each Ideas Roadshow collection offers five separate books presented in an accessible and engaging format.

- *Conversations About Anthropology & Sociology*
- *Conversations About Astrophysics & Cosmology*
- *Conversations About Biology*
- *Conversations About History, Volume 1*
- *Conversations About History, Volume 2*
- *Conversations About History, Volume 3*
- *Conversations About Language & Culture*
- *Conversations About Law*
- *Conversations About Neuroscience*
- *Conversations About Philosophy, Volume 1*
- *Conversations About Philosophy, Volume 2*
- *Conversations About Physics, Volume 1*
- *Conversations About Physics, Volume 2*
- *Conversations About Politics*
- *Conversations About Psychology, Volume 1*
- *Conversations About Psychology, Volume 2*
- *Conversations About Religion*
- *Conversations About Social Psychology*
- *Conversations About The Environment*
- *Conversations About The History of Ideas*

All collections are available as hardcover, paperback and eBook.